The ESSENTIALS® of

Astronomy

D0012524

Charles O. Brass

Instructor of Astronomy
Office of Special Programs
University of Chicago
Chicago, Illinois

Research & Education Association
61 Ethel Road West
Piscataway, New Jersey 08854

Dr. M. Fogiel, Director

THE ESSENTIALS®
OF ASTRONOMY

Year 2004 Printing

Printed in the United States of America

Library of Congress Control Number 98-65424

International Standard Book Number 0-87891-965-1

ESSENTIALS is a registered trademark of Research & Education Association, Piscataway, New Jersey 08854

WHAT "THE ESSENTIALS" WILL DO FOR YOU

This book is a review and study guide. It is comprehensive and it is concise.

It helps in preparing for exams and in doing homework, and remains a handy reference source at all times.

It condenses the vast amount of detail characteristic of the subject matter and summarizes the **essentials** of the field.

It will thus save hours of study and preparation time.

The book provides quick access to the important concepts, definitions, principles, and practices in the field.

Materials needed for exams can be reviewed in summary form—eliminating the need to read and re-read many pages of textbook and class notes. The summaries will even tend to bring detail to mind that had been previously read or noted.

This "ESSENTIALS" book has been prepared by an expert in the field, and has been carefully reviewed to ensure accuracy and maximum usefulness.

Dr. Max Fogiel
Program Director

Contents

Chapter 1
ASTRONOMY—A HISTORICAL PERSPECTIVE

1.1	Astronomy vs. Astrology	1
1.2	Early Sky Observers	1
1.2.1	Aristarchus of Samos (310–230 B.C.E.)	1
1.2.2	Aristotle (384–322 B.C.E.)	2
1.2.3	Hipparchus (Work Contributions from 160–127 B.C.E.)	2
1.2.4	Claudius Ptolemy (Lived Around 140 C.E.)	2
1.2.5	Nicolaus Copernicus (1473–1543)	3
1.2.6	Tycho Brahe (1546–1601)	3
1.2.7	Johannes Kepler (1571–1630)	3
1.2.8	Galileo Galilei (1564–1642)	4
1.2.9	Sir Isaac Newton (1642–1727)	4
1.2.10	Albert Einstein (1879–1955)	4

Chapter 2
SKY BASICS AND CELESTIAL COORDINATE SYSTEMS

2.1	Celestial Sphere	6
2.1.1	Diurnal Motion	7
2.1.2	Celestial Equator	7
2.1.3	Celestial Poles	7
2.1.4	Ecliptic	7
2.1.5	Equinoxes	8
2.1.6	Solstices	8
2.1.7	Observer's Celestial Meridian	8
2.1.8	Hour Circle	9
2.1.9	Hour Angle	10
2.1.10	Local Sidereal Time	10
2.2	Distances Between Celestial Objects	11
2.3	Distance Units to Celestial Objects	11
2.4	Astronomical Unit (AU)	11
2.5	Light Year (LY)	11
2.6	Parsec (pc)	11

2.6.1   Parallax 11
2.7   Determining the Area of the Sky 12
2.8   Celestial Coordinate Systems 13
2.8.1   Horizon Coordinate System.................. 13
2.8.2   Equatorial Coordinate System................ 14
2.8.3   Ecliptic Coordinate System 14
2.8.4   Galactic Coordinate System 14

Chapter 3
TIME RECKONING
3.1   Time Zones 15
3.2   Apparent Solar Time 15
3.3   Mean Solar Time 16
3.4   Standard Time 16
3.5   Local Mean Time.......................... 16
3.6   Universal Time 17

Chapter 4
INSTRUMENTS FOR OBSERVING
4.1   Binoculars 18
4.2   Telescopes 18
4.2.1   Refracting Telescopes 19
4.2.2   Reflecting Telescopes 20
4.2.3   Newtonian Reflector 20
4.2.4   Cassegrain Reflector 20
4.2.5   Ritchey-Chretien Reflector 21
4.2.6   New Generation Telescopes 21
4.3   Catadioptric Telescopes..................... 22
4.4   Non-Optical Range Telescopes 22
4.5   Calculating the Exit Pupil (Beam Width of
        Light Exiting at Eyepiece)................. 23
4.6   Light Gathering Ability of Aperture
        (Objective Lens or Mirror) 23
4.7   Resolving Power of Optical Telescopes 23
4.8   Magnification Power of Optical Telescopes 24

Chapter 5
THE SOLAR SYSTEM

5.1	Model of the Solar System	25
5.2	Origin of the Solar System	25
5.2.1	Gravitational Contraction	26
5.2.2	Condensation	26
5.2.3	Accretion	26

Chapter 6
THE SUN

6.1	Composition, Size, and Brightness	27
6.2	Energy Production	28
6.2.1	Proton-Proton Chain	29
6.2.2	Carbon Cycle	30
6.3	Sun Data Chart	30
6.4	Solar Atmosphere	32
6.4.1	Photosphere	32
6.4.2	Chromosphere	32
6.4.3	Corona	32
6.4.4	Solar Wind	32
6.5	Solar Activity	33
6.5.1	Sunspots	33
6.5.2	Flares	33
6.5.3	Spicules	33
6.5.4	Prominences	33
6.5.5	Faculae and Plages	34
6.6	Sun's Apparent Annual Path	34
6.6.1	Ecliptic	34
6.6.2	Analemma	34
6.6.3	December Solstice	35
6.6.4	June Solstice	35
6.6.5	March Equinox	36
6.6.6	September Equinox	36

Chapter 7
THE PLANETS

7.1	Inferior and Superior Planets	37
7.2	Kepler's Three Laws of Planetary Motion	38
7.2.1	Kepler's First Law	38
7.2.2	Orbits	39
7.2.3	Kepler's Second Law	39
7.2.4	Kepler's Third Law	40
7.2.5	Calculating Surface Gravity	40
7.2.6	Calculating Escape Speed	41
7.3	Terrestrial Planets	41
7.3.1	Mercury	41
7.3.2	Venus	43
7.3.3	Earth	44
7.3.4	Mars	48
7.4	Jovian Planets	50
7.4.1	Jupiter	50
7.4.2	Saturn	51
7.4.3	Uranus	53
7.4.4	Neptune	55
7.4.5	Pluto	56
7.5	Titius-Bode Rule	58

Chapter 8
PLANETARY SATELLITES

8.1	Artificial Satellites	59
8.2	Natural Satellites	59
8.3	The Moon	59
8.4	Moon Phases	60
8.5	Moon Motions	60
8.5.1	Rotation	61
8.5.2	Revolution and Libration	61
8.5.3	Tides	61
8.5.4	Moon Data Chart	62

Chapter 9
ASTEROIDS, METEOROIDS, AND COMETS

9.1   General Information 64
9.1.1   Asteroids 64
9.1.2   Asteroid Belt 64
9.1.3   Meteoroids 65
9.1.4   Meteors 65
9.1.5   Meteorites 65
9.2   Comets 65
9.2.1   Periodic Comets 66
9.2.2   Non-Periodic Comets 66

Chapter 10
ECLIPSES

10.1   Definition 67
10.2   Solar Eclipse 67
10.2.1 Total Solar Eclipse 67
10.2.2 Partial Solar Eclipse 68
10.2.3 Annular Eclipse 68
10.3   Lunar Eclipse 69
10.3.1 Total Lunar Eclipse 69
10.3.2 Partial Lunar Eclipse 69
10.3.3 Penumbral Eclipse 69
10.4   Relationship Between Angular Diameter and Linear
        Diameter (for Small Angles) 70

Chapter 11
STARS

11.1   What is a Star? 71
11.2   Binary Star Systems 72
11.2.1 Optical Double 72
11.2.2 Visual Binary 72
11.2.3 Composite Spectrum Binary 72
11.2.4 Eclipsing Binary 72
11.2.5 Astrometric Binary 73

11.2.6	Spectroscopic Binary	73
11.3	Multiple Star Systems	73
11.4	Variable Stars	73
11.4.1	Cepheid Variables	74
11.4.2	Novae	74
11.4.3	Supernovae	74
11.5	Light	74
11.5.1	Spectra	75
11.6	Magnitudes	75
11.6.1	Absolute Magnitude	76
11.6.2	Apparent Magnitude	76
11.6.3	Mathematical Relationship Between Absolute and Apparent Magnitudes	76
11.7	Stellar Classifications	76
11.7.1	Spectral Classification	77
11.7.2	Luminosity Classification	77
11.8	Hertzsprung-Russel (H-R) Diagram	77
11.9	Determining Stellar Distances	78
11.10	Constellations	79
11.11	Stellar Nomenclature	79
11.12	Stellar Motions	79
11.12.1	Proper Motion	79
11.12.2	Radial Velocity	80
11.12.3	Tangential Velocity	80
11.12.4	Space Velocity	80
11.12.5	Peculiar Velocity	81
11.13	Destiny of Stars	81
11.13.1	White Dwarfs, Red Dwarfs, and Medium-Mass Stars	81
11.13.2	Neutron Stars	82
11.13.3	Black Holes	82
11.14	Masses of Binary Stars	83
11.15	Escape Speed	83
11.16	Relationship Between Frequency and Wavelength (for Electromagnetic Radiation)	83
11.17	Relationship Between Energy and Wavelength (for Electromagnetic Radiation)	84

11.18   Temperature Scales and Conversion Formulas 84
11.19   Intensity Ratio of Two Bodies 85
11.20   Mass-Luminosity Relationship
         (for Main Sequence Stars) 85
11.21   Wien's Law (Used to Obtain the Temperature
         at the Surface of a Star) 85
11.22   Stefan-Boltzman Law (Used to Obtain the
         Energy or Luminosity of a Star) 86
11.23   Luminosity, Radius, and Temperature 86
11.23.1 Alternate Relationship for Luminosity, Radius,
         and Temperature (with Respect to the Sun) 87
11.24   Doppler Shift and Radial Velocity 87
11.25   Life Expectancy of Stars on the Main-Sequence... 88
11.26   Gravitational Red Shift
         (Especially Pertinent to Black Holes) 88
11.27   Schwarzchild Radius 89

Chapter 12
THE OBSERVABLE UNIVERSE

12.1   Boundaries of the Observable Universe 90
12.2   The Big Bang 90
12.2.1   Estimating the Age of the Universe
         Using the Hubble Constant 91
12.3   Galaxies 92
12.4   Star Clusters 93
12.4.1   Open Star Clusters 93
12.4.2   Globular Clusters 93
12.5   Nebulae 94
12.5.1   Absorption Nebula 94
12.5.2   Emission Nebula 94
12.5.3   Reflection Nebula 94
12.6   Quasars 94
12.7   Relativistic Red Shift and Radial Velocity 95
12.8   Hubble's Law 95

Chapter 13
EXTRATERRESTRIAL INTELLIGENCE
13.1   Life Outside of Earth 96

13.2   The Drake Equation 96

CHAPTER 1

Astronomy – A Historical Perspective

1.1 Astronomy vs. Astrology

Astronomy is the scientific study of the Universe and its contents beyond Earth's atmosphere. It should not be confused with astrology, which is a belief that human personality traits are directly influenced by the positions of the Sun, Moon, and planets in relation to the stars. Astrology's contribution to the science of astronomy was the accurate records kept of the positions of the Sun, Moon, and planets with respect to the stars.

1.2 Early Sky Observers

Many early sky observers believed the Universe to be finite in size and almost always placed a stationary Earth at its center. Thus, Earth was also seen as the center of the Solar System. The Sun, Moon, and the five known planets were seen as circling Earth in some fashion and moving against a background of fixed stars.

1.2.1 Aristarchus of Samos (310-230 B.C.E.)

Aristarchus, a Greek astronomer, was the first to propose a Sun-centered (heliocentric) model of the Solar System correctly. He stated that Earth rotated on its axis and revolved around the Sun. This correctly explained the apparent daily motion of the sky and the annual motion of the Sun with respect to the stars. This view was short-lived due to the lack of evidence of stellar parallax and a true understanding of natural physical laws.

1.2.2 Aristotle (384-322 B.C.E.)

Aristotle, the great Greek philosopher and most famous pupil of Plato, devised an Earth-centered (geocentric) model of the Universe based on uniform circular motion. His teachings described Earth as corrupt and changeable and the heavens as perfect and immutable. Although his system did not describe celestial motions very well, his teachings dominated thinking for nearly 1,800 years.

1.2.3 Hipparchus (Work Contributions from 160-127 B.C.E.)

Hipparchus, a Greek astronomer, believed in a geocentric model of the Solar System and adopted two systems to explain naked eye motions. One system involved movable eccentrics (intersecting circles that don't share the same center) and the other involved epicycles (small circles) and deferents (large circles). He is best known for discovering Earth's precessional motion and for laying the foundation for our stellar magnitude scale, which offers a quantitative measure of relative brightness among celestial bodies.

1.2.4 Claudius Ptolemy (Lived Around 140 C.E.)

Ptolemy, a Greek scientist, also believed in a geocentric model of the Solar System. His final model included movable eccentrics, epicycles, deferents, and the equant (devised by him), which is an imaginary point around which an epicycle moves at a uniform rate. This model, along with his table of planetary motions, endured for nearly 15 centuries. Ptolemy compiled a series of 13 volumes on astronomy known as the "Almagest." The Almagest is our main source of Greek astronomy and includes Ptolemy's personal contributions along with a collection of astronomical achievement before his time. The work of Hipparchus is a principal part of this collection.

1.2.5 Nicolaus Copernicus (1473-1543)

Copernicus, a Polish astronomer, believed in a Sun-centered model of the Solar System. He reintroduced the heliocentric concept proposed by Aristarchus of Samos some 18 centuries earlier. His greatest contribution was the concept that Earth is one of six (then known) planets that revolve around the Sun. He arranged these planets in order of increasing distance from the Sun starting with Mercury followed by Venus, Earth, Mars, Jupiter and finally Saturn. This model greatly simplified and correctly explained the phenomenon of retrograde (westward/backwards) motion of the planets without using the cumbersome system of epicycles.

1.2.6 Tycho Brahe (1546-1601)

Tycho, a Danish astronomer, rejected the Copernican model because he couldn't obtain a measurement for stellar parallax. He also rejected the Ptolemaic model because of its inaccurate predictions that grew worse with passing centuries. Tycho's model was a complex Earth-centered system that was short-lived. His great contribution was in providing observational data of unprecedented accuracy.

1.2.7 Johannes Kepler (1571-1630)

Kepler, a German astronomer, worked as an assistant to Tycho Brahe. He accepted the Copernican view but abandoned the 2,000-year-old belief of uniform circular motion. From observational data left by Tycho Brahe, Kepler derived his three laws of planetary motion. Respectively, these three laws are the Law of Elliptical Orbits, the Law of Equal Areas, and the Law of Orbital Periods.

1.2.8 Galileo Galilei (1564-1642)

Galileo, an Italian astronomer, supported a heliocentric view of the Solar System. He was the first person to use the telescope for astronomical observations. He made a series of important discoveries and amassed evidence for support of a Sun-centered system. The four largest moons of Jupiter were first observed by Galileo and are referred to as the Galilean moons.

1.2.9 Sir Isaac Newton (1642-1727)

Newton, a British scientist, accepted a heliocentric view of the Solar System. He examined the mechanics of planetary motion and concluded that gravity was the underlying force acting between orbiting bodies. Realizing this, Newton formulated the Law of Universal Gravitation, which states that the force acting between two bodies is proportional to the product of their masses and inversely proportional to the square of the distance between them.

Symbolically, $F = \dfrac{GMm}{d^2}$

where:

F = gravitational force between the two bodies

G = gravitational constant

M and m = masses of the two bodies

d = distance between the centers of the two bodies

Notice that if M or m increases, then F also increases, and if d increases then F decreases.

1.2.10 Albert Einstein (1879-1955)

In 1905, Einstein, a German-born Jewish physicist, formulated his theory of special relativity, which placed the speed of light (186,282 mi/s) as an upper limit for both Earthly and heavenly material bodies. In 1916, his newly-formulated general theory of relativity revolutionized our concept of gravity by describing it as a curvature of space-time. This profoundly affected our view of cosmology and the entire Universe. Einstein's most famous equation, $E = MC^2$, relates energy to mass and is fundamental in understanding how the Sun and other stars generate such enormous amounts of energy.

CHAPTER 2

Sky Basics and Celestial Coordinate Systems

2.1   Celestial Sphere

The celestial sphere is an imaginary sphere centered on and surrounding Earth. The inside of this hollow sphere is considered as our "model" of the sky upon which the background stars are attached. The Sun, Moon, planets, and other celestial bodies appear to move against this backdrop of seemingly fixed stars. Keep in mind that this is only a two-dimensional model of the sky, as the stars are at varying distances from Earth. The value of this model is that it allows ease in describing positions and motions of astronomical bodies.

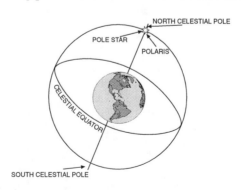

Figure 2.1   The Celestial Sphere

6

2.1.1   Diurnal Motion

The celestial sphere can be imagined as rotating from east to west, thereby making the stars appear to move as such. This apparent westward daily motion of the stars across the sky is actually caused by Earth's eastward rotation and not by any movement of the sky/ celestial sphere.

2.1.2   Celestial Equator

The celestial equator is the projection of Earth's equator onto the sky and celestial sphere. Just as Earth's rotational axis makes an approximate angle of 23½° with the ecliptic, so does the celestial equator.

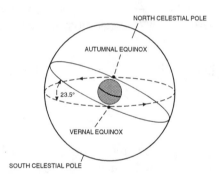

NORTH CELESTIAL POLE

AUTUMNAL EQUINOX

23.5°

VERNAL EQUINOX

SOUTH CELESTIAL POLE

Figure 2.2   Celestial Poles and the Ecliptic

2.1.3   Celestial Poles

The north and south celestial poles are extensions of Earth's rotational axis, which coincide with the north and south geographic poles.

2.1.4   Ecliptic

As viewed from Earth, the ecliptic is the apparent annual path of the Sun along the celestial sphere. As viewed from the Sun, it is the projection of Earth's orbit onto the sky. Solar and lunar eclipses

7

can only occur when the new or full Moon is at or near one of the points at which the Moon's orbit crosses the ecliptic. These points are called nodes.

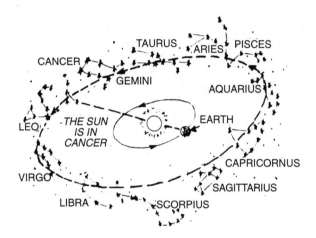

Figure 2.3   The Ecliptic

2.1.5   Equinoxes

These are the two opposite points on the celestial sphere at which the ecliptic crosses the celestial equator. They are called the vernal and autumnal equinoxes. See sections 6.6.5 and 6.6.6.

2.1.6   Solstices

These are the two opposite points on the celestial sphere at which the Sun reaches its greatest declination north ($+23\frac{1}{2}°$) and south ($-23\frac{1}{2}°$) of the celestial equator. They are called the summer and winter solstices, respectively. See sections 6.6.3 and 6.6.4.

2.1.7   Observer's Celestial Meridian

This is the imaginary half-great circle that connects the north and south points on the horizon while passing through the zenith (point directly overhead). The bottom half of this great circle continues below the Earth, through the nadir (point directly below the individual's feet) and returns to the north point.

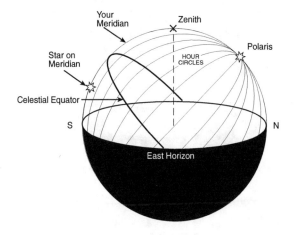

Figure 2.4   The Meridian

2.1.8   Hour Circle

An hour circle is an imaginary north-south half great circle on the celestial sphere that passes through the north and south celestial poles. They are meridians of right ascension.

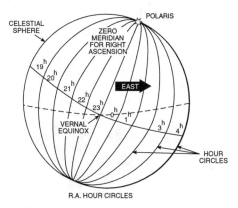

Figure 2.5   Hour Circles

2.1.9   Hour Angle

This is the angular measurement in units of time of how far westward an object is located from the celestial meridian. One hour of time corresponds to 15° of arc.

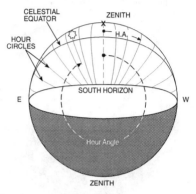

Figure 2.6   Hour Angles

2.1.10   Local Sidereal Time

This is defined as the hour angle of the vernal equinox, which is the right ascension hour circle currently on your meridian. It is time measured with respect to the background stars. See chapter 3 for additional information on time.

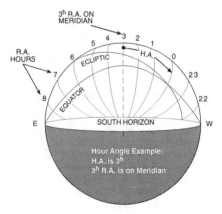

Figure 2.7   Sidereal Time

2.2   Distances Between Celestial Objects

Along the celestial sphere, distances between objects are measured in angular units of degrees, minutes, and seconds.

2.3   Distance Units to Celestial Objects

Distances from Earth to objects in the real sky are measured in the following units:

2.4   Astronomical Unit (AU)

An astronomical unit is the average distance between the Earth and Sun. This is approximately 150 million kilometers, or 93 million miles.

2.5   Light Year (LY)

A light year is the distance light travels through a vacuum in one year. This is approximately 9½ trillion kilometers, or 6 trillion miles.

2.6   Parsec (pc)

A parsec is the distance at which an object would have a parallax of one arcsecond. This is approximately 3.26 light years.

2.6.1   Parallax

Parallax is an apparent displacement or shift of an object due to the motion of an observer. Stellar parallax is the apparent displacement of a nearby star that results from the motion of the Earth around the Sun. Numerically, this is the angle bounded by 1 AU at the distance of any given star (see figure 2.8).

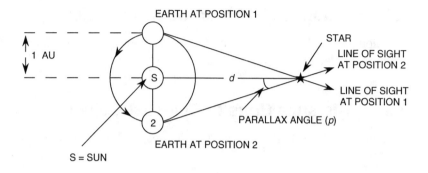

Figure 2.8 Positions During Parallax

The distance (d) in parsecs equals the inverse of the parallax angle (p) in arcseconds.

Symbolically, $d = \dfrac{1}{p}$

where:

d = the distance between the Sun and star in parsecs

p = the parallax angle in arcseconds

Notice that the larger the parallax angle, the closer the object.

2.7   Determining the Area of the Sky

We can determine the angular area of the sky (celestial sphere) by recalling that the area, A, of a sphere is $4\pi r^2$ and the circumference, C, of a circle is $2\pi r$.

$$A = 4\pi r^2 \tag{1}$$

$$C = 2\pi r \tag{2}$$

Then, $r = \dfrac{c}{2\pi}$ or $r = \dfrac{360°}{2\pi}$ $\tag{3}$

Substituting Equation 3 in Equation 1:

$$A = 4\pi \left(\frac{360°}{2\pi} \right)^2$$

$$= \frac{4\pi}{4\pi^2} (129,600 \ deg^2)$$

$$= \frac{129,600 \ deg^2}{\pi}$$

$$= 41,252.96 \ deg^2$$

This is the total area of the sky surrounding Earth. At any given time, we can only see one-half of the sky, as the other half is below our line of sight. Therefore, the area of sky that's visible to an observer at any given time is one-half of the total sky area or approximately 20,626.48 square degrees.

2.8 Celestial Coordinate Systems

Celestial coordinate systems are used in specifying positions of celestial objects or points in the sky. Use of a system is based upon convenience. For example, when studying the planets, it is convenient to work with the ecliptic coordinate system as the planets never stray far away from the "ecliptic" reference frame.

2.8.1 Horizon Coordinate System

This system specifies positions in the sky using angular coordinates called azimuth and altitude.

Azimuth measures angular distance around your horizon clockwise (from 0 to 360°), starting at the north point.

Altitude measures angular distance above the horizon (0°) to your zenith (90°).

2.8.2 Equatorial Coordinate System

This system specifies positions in the sky using time and angular coordinates called right ascension and declination, respectively.

Right Ascension (RA) measures angular direction in units of time (0 to 24 hours) eastward along the celestial equator from the vernal equinox. It is convenient to use time, because it relates the position of a star to its apparent motion across the sky.

Declination (Dec) measures angular direction in degrees north (+) and south (–) from the celestial equator (at $0°$) to the celestial poles (at $±90°$).

2.8.3 Ecliptic Coordinate System

This system specifies positions in the sky using angular coordinates called celestial longitude and celestial latitude.

Celestial Longitude is measured eastward along the ecliptic (from $0 – 360°$), starting at the vernal equinox.

Celestial Latitude measures positions in degrees north (+) and south (–) from the ecliptic (at $0°$) to the ecliptic poles (at $±90°$).

2.8.4 Galactic Coordinate System

This system specifies positions within our galaxy using angular coordinates called galactic longitude and galactic latitude.

Galactic Longitude is measured eastward along the galactic equator (from $0 – 360°$), starting at an imaginary line that connects the Sun with the center of our galaxy.

Galactic Latitude measures positions in degrees north (+) and south (–) from the galactic equator (at $0°$) to the galactic poles (at $±90°$).

CHAPTER 3

Time Reckoning

3.1   Time Zones

Time zones represent the 24 different timekeeping areas into which the Earth is divided. Central meridian lines of longitude, at 15° intervals, separate these successive time zones. Since Earth completes one rotation (360°) in one day (approximately 24 hours), 15° of arc is equivalent to one hour of time. The zero degree reference meridian is centered on Greenwich, England. Moving 15 degrees east of this meridian adds one hour to your time and moving 15 degrees west subtracts one hour. The International Date Line (date change line) is located directly opposite the Greenwich meridian at 180°. Crossing the International Date Line going west necessitates advancing your time by one day; going east necessitates delaying your time by one day.

3.2   Apparent Solar Time

Apparent solar time or apparent time is time reckoned with respect to the motion of the real Sun. It is the time shown by a sundial. This is not a very practical way of keeping time since the motion of the real Sun is not uniform. Also, very close east and west neighboring locations would show a slightly different solar time at the same instant. To locate the exact position of the Sun, you must use apparent solar time.

3.3　Mean Solar Time

Mean solar time, or, simply, mean time is time reckoned with respect to a fictitious Sun. This fictitious Sun travels at a steady speed equal to the average speed of the real Sun. It is the time shown by your watch. Mean solar time (clock time) and apparent solar time (sundial time) can differ by as much as 16 minutes. This time difference is known as the equation of time and results from the combined effects of Earth's axial tilt and elliptical orbit.

3.4　Standard Time

Standard time is time reckoned with respect to the central (standard) meridian of a specific time zone. It is the everyday clock time used within each one hour time zone. For example 9:00 p.m. standard time indicates the clock time used for all locations within 7½° on both sides of the central meridian. This kind of time is not very helpful in trying to locate the exact positions of celestial objects.

3.5　Local Mean Time

Local mean time is the "specific" time at a particular location rather than the general "standard time" that's normally used. It is also called local civil time and is a consequence of the natural variation of time with longitude. The variation in local mean time is four minutes of time for every degree of arc east or west. For example, when it's 9:00 p.m. on the central meridian, it's 9:00 p.m. **standard time** for all locations within 7½° of this meridian. However, this is not the case for the local mean time. One degree to the east of the central meridian the local mean time is four minutes later (i.e., 9:04 p.m.); one degree to the west it is four minutes earlier (i.e., 8:56 p.m.). Except for the Sun, you must use local mean time to locate all celestial bodies in the sky.

3.6   Universal Time

Universal time is time ascertained with respect to the zero meridian. It is the same as Greenwich mean time. Traditionally, astronomers used the term Universal Time or U.T. and navigators used Greenwich mean time or G.M.T. Converting between Universal time and standard time is quite simple. Use the following formulas.

From Standard Time to Universal Time:

U.T. = Standard Time + the zone number

From Universal Time to Standard Time:

Standard Time = U.T. − the zone number

The following is a list of zone numbers within the continental U.S.A.:

Zone +4 = Atlantic Standard Time

Zone +5 = Eastern Standard Time

Zone +6 = Central Standard Time

Zone +7 = Mountain Standard Time

Zone +8 = Pacific Standard Time

CHAPTER 4

Instruments for Observing

4.1 Binoculars

Binoculars are essentially two small telescopes mounted parallel to one another, thus allowing both eyes to view an object. The optics in these precision instruments include prisms that provide an upright image. As with all instruments used in astronomical observation, the two main functions are 1) to gather more light than the human eye can gather and 2) to magnify the image. All binoculars are labeled with a pair of numbers that indicate magnification power and objective lens size (in millimeters) respectively. For example, a pair of 10 × 70 binoculars has a magnification power of ten times and an objective lens size of 70 mm. In agreement with the maximum dilation of the human eye, binoculars having an exit pupil of 7 millimeters are best for night observation. Because they provide a wider field of view than telescopes, binoculars are excellent for comet hunting and Moon observation.

4.2 Telescopes

A telescope is used to collect electromagnetic radiation (e.g., visible light and radio waves) from afar, focus it and then magnify the image or signal. For optical telescopes, the objective lens acts as the light gathering element and the ocular (or eyepiece) acts as the image magnifier. Magnification is not so important because the telescope eyepiece can be changed quite readily for the desired magnification. Telescopes are recognized by the size of their objective lens or mirror.

Unlike terrestrial telescopes, astronomical telescopes produce an inverted image of the object in view. Additional optics could upright this image at the cost of interfering with an already dim amount of light. The diameter of the aperture (main lens or mirror) is the most important feature for any optical telescope. Light gathering ability and resolving power are increased as aperture size increases.

Increased light gathering ability allows fainter objects to be seen and produces brighter images. Resolving power allows the instrument to separate objects that lie close together and to see liner details.

4.2.1   Refracting Telescopes

A refracting telescope has a main or primary (objective) lens as its light gathering element. Incoming light is refracted (bent) by the objective lens and forms an image at the eyepiece where it is magnified.

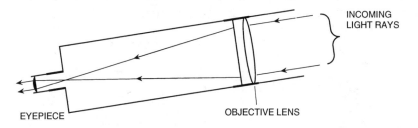

INCOMING
LIGHT RAYS

EYEPIECE   OBJECTIVE LENS

Figure 4.1   Refracting Telescope

The world's largest refracting telescope is the 40 inch (1 meter) refractor at Yerkes Observatory in Williams Bay, Wisconsin. Engineering difficulties arise when trying to construct refractors with increasingly larger objective lenses. All of the largest telescopes are reflectors, as production is easier and cheaper and the engineering difficulties that arise when constructing refractors are avoided. Spherical aberration and chromatic aberration are two types of focusing difficulties encountered with refractors.

4.2.2   Reflecting Telescopes

A reflecting telescope has a main or primary (objective) mirror as its light gathering element. This concave mirror collects and focuses incoming light to form an image. Because the image is formed in front of the mirror. In the path of incipient light, astronomers sit inside of a cage at the point called the prime focus where they view a magnified image through the eyepiece. The world's largest single-mirror reflecting telescope is the 236-inch (6 meter) reflector near Zelenchukskaya in the Crimea (former Soviet Union). The second largest is the 200-inch (5 meter) Hale telescope on Mount Palomar near Pasadena, California.

4.2.3   Newtonian Reflector

A newtonian reflector uses a flat mirror or prism to redirect the light from the primary mirror before the image is formed. This light is reflected to the eyepiece located at the side of the telescope tube. Sir Isaac Newton devised this type of system, and instruments employing such optics are called Newtonian reflectors.

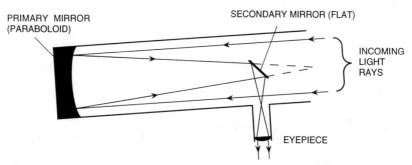

PRIMARY MIRROR
(PARABOLOID)

SECONDARY MIRROR (FLAT)

INCOMING
LIGHT
RAYS

EYEPIECE

Figure 4.2   Newtonian Reflector

4.2.4   Cassegrain Reflector

A cassegrain reflector uses a convex (curved outwards) mirror to redirect the light from the prime focus back through a hole at the center of the primary mirror. This type of telescope has a paraboloid as its primary mirror and a convex hyperboloid as its secondary mirror. In contrast, Gregorian reflectors have a secondary mirror that is concave (curved inwards). Unlike refractors, reflectors never suffer with chromatic aberration but are subject to spherical aberration unless their primary mirror is a paraboloid. Spherical aberration was the problem found with the primary mirror of the Hubble space telescope.

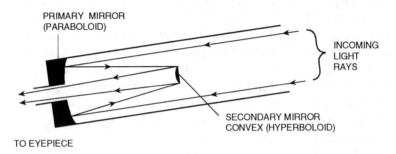

Figure 4.3   Cassegrain Reflector

4.2.5   Ritchey-Chretien Reflector

The Hubble Space Telescope (HST) is a Ritchey-Chretien reflector and was placed in Earth orbit on April 25, 1990, by the space shuttle *Discovery*. Earth's atmosphere is opaque to various wavelengths of electromagnetic radiation and it also causes blurred images. The HST is 43 feet in length, and is equipped with a primary mirror that is 94.5 inches (2.4 meters) in diameter. In late 1993, a space shuttle servicing crew successfully installed a corrective optics package that rectified the problem with the primary mirror. (Subsequently, the Hubble began beaming back outstanding images of a vast array of cosmic phenomena.) This type of reflector has a cassegrain focusing system and employs a hyperboloid primary and secondary mirror. Reflectors having this kind of optical system have the advantage of offering a wide field of view.

4.2.6   New-Generation Telescopes

Segmental mirror telescopes are reflecting telescopes that use an array of individual mirrors designed and mounted to collect light for focusing. The Keck telescope, located on Mauna Kea in Hawaii, is a segmental mirror telescope employing this technique. It has an array of 36 hexagonal mirrors with the light gathering ability equivalent to a 394-inch (10 meter) diameter reflector. This instrument qualifies as the largest reflecting (and optical) telescope in the world. A multiple-mirror telescope has a somewhat similar design. The multimirror instrument at the Whipple observatory on Mount Hopkins near Tucson, Arizona, consists of six 72-inch (1.8 meter) mirrors in a circular array having the equivalent light gathering ability of a single 176-inch (4.5 meter) diameter reflector. Such instruments are made possible by having the capability of producing thinner lightweight mirrors and computer support that interfaces with the optics system. The production of single large diameter (e.g., 8 meters) lightweight thinnet mirrors are part of an even newer trend.

4.3   Catadioptric Telescopes

A catadioptric telescope uses a combination of lenses and mirrors in its optics design. The 48-inch (1.2 meter) Schmidt telescope on Mount Palomar is a notable example.

4.4   Non-Optical Range Telescopes

Non-optical telescopes are designed to collect electromagnetic radiation outside of the visible spectrum. These ranges of frequencies are necessary for study because visible light is not the only wavelength at which objects radiate. Furthermore, some celestial objects emit no visible light at all. Non-visible wavelengths probed include, radio, infrared, ultraviolet, X-ray and gamma-ray. The Hubble space telescope operates in both optical and non-optical ranges.

4.5 Calculating the Exit Pupil (Beam Width of Light Exiting at Eyepiece)

$$\text{Exit Pupil} = \frac{\text{Diameter of Objective Lens (in millimeters)}}{\text{Magnification Power of Eyepiece}}$$

For maximum utilization of light gathered by the objective lens or mirror, the beam width should correspond to the maximum diameter that the pupil of the human eye can dilate. This decreases with age (and health issues) from approximately 9 mm to 4 mm.

4.6 Light Gathering Ability of Aperture (Objective Lens or Mirror)

$$\frac{L_1}{L_2} = \left(\frac{D_1}{D_2}\right)^2$$

where:

L_1 = Light-gathering ability of aperture 1

L_2 = Light-gathering ability of aperture 2

D_1 = Diameter of aperture 1

D_2 = Diameter of aperture 2

4.7 Resolving Power of Optical Telescopes

$$\propto = \frac{\lambda}{d}(206,265)$$

where:

$\propto$ = Resolution (in radians)

λ = Wavelength (in meters)

d = Diameter of aperture (in meters)

This is the theoretical resolving power. In reality the angular resolution of ground-based optical telescopes is limited by the quality of seeing.

4.8   Magnification Power of Optical Telescopes

$$M = \frac{F_o}{F_e}$$

where:

M = Magnification power

F_o = Focal length of objective (in centimeters)

F_e = Focal length of eyepiece (in centimeters)

CHAPTER 5

The Solar System

5.1 Model of the Solar System

The Solar System consists of the Sun, nine major planets, natural planetary satellites, asteroids, meteoroids, comets, and the interplanetary medium. At the center of this system is the Sun, which holds all members captive within its immense gravitational field. All members of the Solar System are in different orbits around the Sun and traveling at various speeds and distances.

5.2 Origin of the Solar System

German philosopher Immanuel Kant (1755) and French astronomer Pierre Simon de Laplace (1796), respectively proposed and developed the idea that the Solar System formed from a rotating flattened gaseous cloud, later called the solar nebula. This idea, known as the Nebular Hypothesis, proposes that the Sun and the planets formed from a rapidly rotating interstellar cloud of gas and dust. Gravitational contraction caused continued collapse of this cloud, with conservation of angular momentum resulting in increased rotation and a flattened disk with a central bulge. Approximately 4.5 billion years ago, the Sun, planets, and other members of the Solar System formed from this configuration through the processes of gravitational contraction, condensation, and accretion.

5.2.1 Gravitational Contraction

Gravitational contraction is the contraction of a body due to self-gravitation. It is caused by the mutual gravitational attraction between the masses of particles within a body.

5.2.2 Condensation

Condensation is the condensing of solid materials from gaseous ones. It is a temperature-dependent process and accounts for the differences in chemical composition among planets.

5.2.3 Accretion

Accretion is the coalescence of smaller particles into larger ones. It is accomplished through either mutual gravitational attraction or as a result of chance collisions. The coalescing of planetesimals (small bodies) forms protoplanets (larger bodies destined to become planets).

CHAPTER 6

The Sun

6.1   Composition, Size, and Brightness

The Sun is a neutrally-charged, hot glowing body of ionized gas, consisting of free electrons and atomic nuclei. Matter in this form is called plasma and is referred to as the fourth state of matter. It is estimated that over 99 percent of the matter in the Universe, including most bright stars, exists in this form.

Our Sun is a star of medium size and brightness when compared with others in our galaxy. When plotted on the Hertzsprung-Russel diagram (i.e., a plot of luminosity vs. temperature), the Sun, along with approximately 90 percent of all stars, lies along a narrow band called the "main sequence." This is actually a "mass" sequence of stars, and the Sun lies midway along the sequence, which puts it among medium-sized stars. Although the vast majority of stars are fainter than the Sun, its absolute magnitude of approximately +5 centers close around the midpoint of the range from the faintest stars (+20) to the brightest (–9).

That is, the difference between +5 and –9 is +5 – (–9) which is 14 magnitudes; the difference between +5 and +20 is +5 – (+20) which is 15 magnitudes thus placing the Sun around the midpoint range of brightness.

6.2   Energy Production

The Sun generates energy by thermonuclear fusion. Each second, 600 million tons plus of hydrogen (protons) are fused into helium with the simultaneous release of enormous amounts of energy. This energy comes from converting a small fraction of the hydrogen mass into energy and can be calculated from Einstein's famous equation

$$E = MC^2$$

where E is energy, M is mass, and C is the speed of light in free space.

The fractional amount of hydrogen available for conversion can be determined as follows:

Given:

1. It takes four ordinary hydrogen ($_1^1\text{H}$) atoms to make one ordinary helium ($_2^4\text{He}$) atom.

2. The mass of an ordinary hydrogen atom is 1.007825 amu (atomic mass units).

3. The mass of an ordinary helium atom is 4.00268 amu.

Therefore:

4. Four hydrogen atoms have a mass of 4 × (1.007825 amu) = 4.03130 amu.

5. The difference in mass between four hydrogen atoms and one helium atom (i.e., the mass defect) is 4.03130 – 4.00268 = 0.02862 amu. The resulting percent change in mass, 0.02862/4.03130 = 0.71%, is the fractional amount of hydrogen actually converted into energy. This amounts to approximately five million tons of hydrogen mass being converted to energy each second. At this rate of conversion, our Sun, as we currently know it, can last for another five billion years. Thus, the hydrogen burning phase of the Sun is approximately ten billion years.

6.2.1   Proton-Proton Chain

The fusion of hydrogen into helium is accomplished through two different nuclear reactions; the proton-proton (P-P1) chain and the carbon-nitrogen-oxygen cycle (CNO or carbon cycle for short). The proton-proton chain is dominant in stars like the Sun whose core temperatures are below 15 million Kelvin.

Of the three possible ways for completing the proton-proton chain, the following reaction (PP 1) occurs 91 percent of the time:

$$^1_1H + {^1_1}H \rightarrow {^2_1}D + e^+ + \nu \qquad \text{releasing 1.442 MeV of energy}$$

$$^2_1D + {^1_1}H \rightarrow {^3_2}He + \gamma \qquad \text{releasing 5.494 MeV of energy}$$

Repeating the above steps to obtain an additional light weight helium atom (^3_2He), we have:

$$^1_1H + {^1_1}H \rightarrow {^2_1}D + e^+ + \nu \qquad \text{releasing 1.442 MeV of energy}$$

$$^2_1D + {^1_1}H \rightarrow {^3_2}He + \gamma \qquad \text{releasing 5.494 MeV of energy}$$

Now combining the two lightweight helium atoms, we get:

$$^3_2He + {^3_2}He \rightarrow {^4_2}He + 2{^1_1}H \qquad \text{releasing 12.859 MeV of energy}$$

Since six protons are involved in completing this chain reaction and two are left over, only four protons actually went into forming the ordinary helium atom. Thus, the foregoing chain reaction may be summarized as follows:

$$4\left(^1_1H + e^-\right) \rightarrow {^4_2}He + 2e^- + 26.73 \text{ MeV of energy}$$

or

4.28×10^{-5} ergs of energy

Note: Electrons are involved in this process as mass and charge must be conserved.

6.2.2   Carbon Cycle

Above 15 million K, the fusion of hydrogen into helium is dominated by the carbon-nitrogen-oxygen (CNO) cycle (or carbon cycle for short). In the Sun, this cycle accounts for a small percentage of the nuclear energy-producing reactions. Its chain proceeds as follows:

$$^{12}_{6}C + ^{1}_{1}H \rightarrow ^{13}_{7}N + \gamma \qquad \text{releasing 1.943 MeV of energy}$$

$$^{13}_{7}N \rightarrow ^{13}_{6}C + e^{+} + \nu \qquad \text{releasing 2.221 MeV of energy}$$

$$^{13}_{6}C + ^{1}_{1}H \rightarrow ^{14}_{7}N + \gamma \qquad \text{releasing 7.551 MeV of energy}$$

$$^{14}_{7}N + ^{1}_{1}H \rightarrow ^{15}_{8}O + \gamma \qquad \text{releasing 7.297 MeV of energy}$$

$$^{15}_{8}O \rightarrow ^{15}_{7}N + e^{+} + \nu \qquad \text{releasing 2.753 MeV of energy}$$

$$^{15}_{7}N + ^{1}_{1}H \rightarrow ^{12}_{6}C + ^{4}_{2}He \qquad \text{releasing 4.966 MeV of energy}$$

As in the P-P1 chain, the net result of the CNO cycle is the conversion of four protons into helium with an energy release of 26.73 MeV.

6.3   Sun Data Chart

Approximate Age	4.6 billion yrs.
Equatorial Diameter	1.39×10^{6} km or
	863,746 mi
Angular Diameter	31.4 to 32.5 arcminutes
Mass	1.99×10^{30} kg or
	332,946 Earth masses
Volume	1,333,000 Earths
Average Density	1.41 g/cm³
Maximum Distance from Earth	1.521×10^{8} km or
	9.451×10^{7} mi

Minimum Distance from Earth	1.471×10^8 km or
	9.142×10^7 mi
Average Distance from Earth	1.496×10^8 km or
	9.296×10^7 mi
Average Surface Temperature	5,800° or
	9,980° F
Core Temperature	1.5×10^{7}° or
	2.7×10^{7}° F
Luminosity	3.83×10^{26} Joules/sec
Apparent Magnitude	–26.7
Absolute Magnitude	+4.8
Surface Gravity	27.9 times Earth
Surface Escape Velocity	618 km/s or
	384 mi/s
Rotational Period	25 Earth days at equator
(at equator)	
Galactic Orbital Period	220 million yrs.
Galactic Orbital Speed	250 km/s or
	155 mi/s
Inclination of Equator to Ecliptic	7.2°
Magnetic Field (at sunspots)	100 to 4,000 Gauss
Spectral Type	G2 V

6.4   Solar Atmosphere

Three regions of solar atmosphere are identified: the photosphere (visible surface), chromosphere (middle layer above photosphere), and the corona (outermost layer).

6.4.1   Photosphere

Referred to as the Sun's surface, this visible gaseous layer is responsible for emitting the sunlight we actually see. It is marked by cellular patterns called granulations that are caused by hot rising gases. Temperature in this region decreases with increasing height.

6.4.2   Chromosphere

Meaning "sphere of color," the pinkish glow of the chromosphere can be directly observed only during a total solar eclipse. Its temperature slowly rises with increasing height.

6.4.3   Corona

This faint, whitish, gaseous halo can be directly observed only during a total solar eclipse. Its shape is more rounded at sunspot maximum. The brighter inner region is called the "K" corona; and the fainter outer region is called the corona. The temperature in this atmospheric region rises with increasing height, reaching values as high as two million Kelvin.

6.4.4   Solar Wind

The solar wind is the continuous outward flow of particles (primarily protons and electrons) from the Sun caused by the very high outward pressure in the corona. These particles range in speed from 200 km/s to 1,000 km/s (124 mi/s to 621 mi/s) and are part of the interplanetary medium.

6.5 Solar Activity

Energetic solar phenomena occurring within the Sun's various atmospheric layers are called solar activity. This activity includes sunspots, flares, spicules, prominences, plages, and faculae.

6.5.1 Sunspots

Sunspots are photospheric regions in which the temperature is lower than in the surrounding areas. This temperature difference, caused by strong magnetic fields, causes sunspots to appear visually darker. Sunspot numbers are associated with an average cycle of approximately 11 years.

6.5.2 Flares

Flares are sudden outbursts of energy from the chromosphere and corona that typically peak over several minutes and slowly fade within an hour. They are associated with sunspots and interfere with radio communications and cause auroral displays on Earth.

6.5.3 Spicules

Spicules are vertical jets of gas emanating from the chromosphere and having a duration of five to ten minutes. Typically, they are 1,000 kilometers across and 10,000 kilometers high.

6.5.4 Prominences

Prominences are cloud- and flamelike structures occurring in the chromosphere and corona. They are less violent than flares and, when viewed along the Sun's edge, appear as bright features in the corona. Viewed against the photosphere, they appear as long dark markings called filaments. They can appear as sprays, loops, or arches.

6.5.5   Faculae and Plages

Faculae are bright regions on the photosphere that appear in the same areas as subsequent sunspots. They also coincide with plages, which are bright regions appearing in the above chromosphere. Plages were formerly known as flocculi.

6.6   Sun's Apparent Annual Path

As the Sun makes its annual journey across the sky, it traces out an unchanging pattern with points along its pathway corresponding to specific times of the year. The ecliptic, analemma, solstices, and equinoxes indicate such paths and points related to this annual motion.

6.6.1   Ecliptic

As viewed from Earth, the ecliptic is the apparent annual path of the Sun along the celestial sphere. As viewed from the Sun, it is the projection of Earth's orbit onto the sky. Solar and lunar eclipses can only occur when the new or full Moon is at or near one of the points at which the Moon's orbit crosses the ecliptic. These points are called nodes.

6.6.2   Analemma

Photographing the Sun's position in the sky at the same time of day over the course of a year results in a composite figure eight pattern known as the analemma. This elongated figure eight pattern is sometimes seen on globes of the Earth. It crosses Earth's equator extending $23\frac{1}{2}°$ to the north and south. Points along its path show the Sun's declination (latitude) throughout the year. It also shows the difference between apparent solar time and mean solar time for any day of the year.

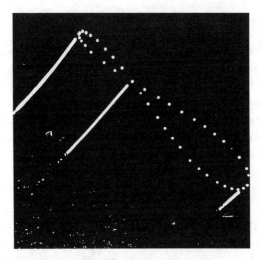

Figure 6.1   The Pattern of the Sun: The Analemma

6.6.3   December Solstice

This is the point on the celestial sphere at which the Sun reaches its greatest declination south (23½°) along the ecliptic with respect to the celestial equator. It marks the beginning of winter for the Northern Hemisphere and is commonly referred to as the winter solstice. The bottom point of the analemma corresponds with this time of year, which is on or about December 22. This is also the shortest day of the year.

6.6.4   June Solstice

This is the point on the celestial sphere at which the Sun reaches its greatest declination north (23½°) along the ecliptic with respect to the celestial equator. It marks the beginning of summer for the Northern Hemisphere and is commonly referred to as the summer solstice. The top point on the analemma corresponds with this time of year, which is on or about June 21. This is also the longest day of the year.

6.6.5   March Equinox

This is the point (0°) on the celestial sphere at which the Sun crosses the celestial equator as it journeys northward along the ecliptic. It marks the beginning of spring for the Northern Hemisphere and is commonly referred to as the spring or vernal equinox. The intersection of the analemma with the equator corresponds with this time of year, which is on or about March 21. Theoretically, day and night are of equal length throughout the world.

6.6.6   September Equinox

This is the point (0 degrees) on the celestial sphere at which the Sun crosses the celestial equator as it journeys southward along the ecliptic. It marks the beginning of autumn for the Northern Hemisphere and is commonly referred to as the fall or autumnal equinox. The intersection of the analemma with the equator corresponds with this time of year, which is on or about September 22. Theoretically, day and night are of equal length throughout the world.

CHAPTER 7

The Planets

7.1 Inferior and Superior Planets

Planets are astronomical bodies that orbit the Sun or another star. They cannot produce their own light and therefore shine by reflecting starlight. Planets orbiting closer to the Sun than Earth are called "inferior" planets (i.e., Mercury and Venus). Those that orbit farther away than Earth are called "superior" planets (i.e., Mars, Jupiter, Saturn, Uranus, Neptune, and Pluto).

Four basic alignment positions of planets with reference to Earth and Sun are recognized. When an inferior planet lies on the far side of the Sun as seen from Earth, it is at "superior conjunction" and rises with the Sun. It continues on to its maximum eastern angular separation called "greatest eastern elongation" and on to "inferior conjunction" where it lies directly between Earth and Sun while once again rising with the Sun. Finally it arrives at its maximum western angular separation called "greatest western elongation" and the cycle repeats. Because elongation is restricted for inferior planets, Mercury can never be more than 28° (1 hour, 52 min.) from the Sun and Venus never more than 47° (3 hrs., 8 min.). This means that these two planets are always in the daytime sky.

The four alignment positions for the superior planets are as follows:

1) Conjunction is when a superior lies on the far side of the Sun as seen from Earth and rises with the Sun.

2) Eastern quadrature is when the planet is 90° east (6 hours behind) the Sun.

3) Opposition is when Earth lies between the Sun and a planet thereby making the planet opposite the Sun by 180° or 12 hours.

4) Western quadrature is when a planet is 90° west (6 hours ahead) of the Sun.

7.2 Kepler's Three Laws of Planetary Motion

Kepler's laws of planetary motion are three fundamental laws that govern the movement of planets as they orbit the Sun. In numerical order respectively, they are the Law of Elliptical Orbits, the Law of Equal Areas, and the Law of Orbital Periods.

7.2.1 Kepler's First Law

All planets revolve around the common center of mass of their Sun-planet system along closed orbits called ellipses, with the Sun at one focus.

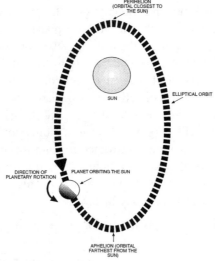

Figure 7.1 Planetary Orbit

7.2.2   Orbits

The trajectory of an orbit is either an open or closed conic section. A body in a continuous orbit around another (e.g., a planet or periodic comet around the Sun) travels in a closed orbit described as an ellipse. An ellipse is a circle with an eccentricity (flatness). A circle is an example of a special-case ellipse and has an eccentricity of "exactly" zero—indicating that it is "perfectly" round. In actual practice, orbits described as circular are only circular out to some order of approximation. However, due to practical limitations in assessing measurements, these nearly circular orbits can be indiscernible from truly circular orbits.

A body that moves along an open orbit has a one time encounter with another body as it travels in an orbit described as hyperbolic. Non-periodic comets are examples of bodies having hyperbolic orbits. A parabola is a one-branch hyperbola having an eccentricity of "exactly" one and whose sides become increasingly steeper. Hyperbolas with eccentricities greater than one have two distinct symmetrical branches but only one of these branches is needed in describing an orbital path. The sides of this hyperbola continually decrease in steepness.

7.2.3   Kepler's Second Law

The rotating position vector of a planet moves across equal areas of its orbit in equal intervals of time.

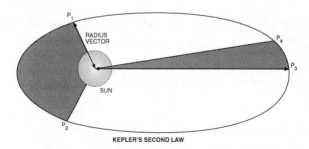

KEPLER'S SECOND LAW

Figure 7.2   View of Kepler's Second Law

7.2.4   Kepler's Third Law

The square of a planet's orbital period (P) is directly proportional to the cube of its semimajor axis (a).

Therefore, Kepler's Third Law of Planetary Motion is

$$P^2 = a^3 \frac{4\pi^2}{G(M_S + M_P)} \approx \frac{A^3}{M_S}$$

where:

P = Orbital period of planet (in years)

a = Average distance of separation (in A.U.'s)

M_S = Mass of sum (in solar masses) Note: $M_S + M_P \approx M_S$

M_P = Mass of planet

G = Gravitational constant

$\dfrac{4\pi^2}{G}$ = Constant = 1 (in Sun-Earth unit system)

7.2.5   Calculating Surface Gravity

$$g = \frac{GM}{R^2}$$

where:

g = Surface gravity (acceleration due to gravity at surface) in meters/sec^2

G = Gravitational constant $\left(6.67 \times 10^{-11} \dfrac{N \times m^2}{kg^2} \right)$

M = Mass of body (in kilograms)

R = Radius of body (in meters)

7.2.6   Calculating Escape Speed

$$V_{esc} = \sqrt{\frac{2GM}{R}}$$

where:

V_{exc} = Surface of escape speed (in meters/sec)

G = Gravitational constant $\left(6.67 \times 10^{-11} \, \dfrac{N \times m^2}{kg^2} \right)$

M = Mass of body (in kilograms)

R = Radius of body (in meters)

7.3   Terrestrial Planets

With respect to their increasing distance from the Sun: Mercury, Venus, Earth, and Mars are the innermost planets of the Solar System and are known as the terrestrial, or Earth-like planets. All are rocky in nature and have a similar fundamental structure.

7.3.1   Mercury

Mercury is the closest planet to the Sun and can be seen with the unaided eye just before sunrise or just after sunset. It exhibits phases similar to the Moon which can be observed through a small telescope. Because it leads or follows the Sun so closely, it is often difficult to spot. In order of relative abundance, its tenuous atmosphere consists of sodium, potassium, helium, hydrogen, and trace gases. Most of its surface has been heavily cratered by meteoric impact. Mercury has an iron core that represents 80 percent of its mass.

MERCURY DATA CHART

Mass   5.6% of Earth or

                                      3.31×10^{23} kg

Equatorial Diameter	4,847 km or
	3,012 mi
Angular Diameter	10.9 arcseconds
(at closest approach)	
Volume	5.6% of Earth
Average Density	5.4/cm^3
Equatorial Surface Gravity	38% of Earth
Equatorial Escape Speed	4.3 km/s or
	2.7 mi/s
Average Surface Temperature	−270° F to 800° F
Natural Satellites	0
Ring System	none
Maximum Apparent Magnitude	−1.9
Average Albedo	0.106
Maximum Distance from Sun	6.97 × 10^7 km or
	4.33 × 10^7 mi
Minimum Distance from Sun	4.59 × 10^7 km or
	2.85 × 10^7 mi
Average Distance from Sun	5.79 × 10^7 km or
	3.60 × 10^7 mi
Sidereal Rotational Period	58.6 days
Oblateness	0
Sidereal Orbital Period	87.97 days
Average Orbital Speed	47.9 km/s or
	29.8 mi/s
Orbital Eccentricity	0.2056

Axial Tilt or Obliquity	0°
(inclination of equator to orbit)	
Orbital Tilt	7°
(inclination of orbit to ecliptic)	
Magnetic Field (measured)	3.3×10^{-3} Gauss

7.3.2 Venus

Venus is the second planet from the Sun, and it too can be seen with the unaided eye just before sunrise or just after sunset. It also exhibits phases similar to the Moon. The thick atmosphere results in an enormous surface pressure equal to 90 times that of Earth. Venus has the highest surface temperature of all the planets due to the greenhouse effect, caused by its atmosphere being composed primarily of carbon dioxide (96 percent by volume) and clouds containing sulfuric acid. Venus has a semisolid core of iron and nickel.

VENUS DATA CHART

Mass	82% of Earth or
	4.87×10^{24} kg
Equatorial Diameter	12,104 km or
	7,521 mi
Angular Diameter	61 arcseconds
(at closest approach)	
Volume	86% of Earth
Average Density	5.24 g/cm^3
Equatorial Surface Gravity	91% of Earth
Equatorial Escape Speed	10.3 km/s or
	6.4 mi/s

Surface Temperature	882° F
Natural Satellites	0
Ring System	none
Maximum Apparent Magnitude	−4.4
Average Albedo (cloud tops)	0.76
Maximum Distance from Sun	1.089×10^8 km or
	6.767×10^7 mi
Minimum Distance from Sun	1.075×10^8 km or
	6.680×10^7 mi
Average Distance from Sun	1.082×10^8 km or
	6.724×10^7 mi
Sidereal Rotational Period	243 days retrograde
Oblateness	0
Sidereal Orbital Period	224.7 days
Average Orbital Speed	35 km/s or
	21.8 mi/s
Orbital Eccentricity	0.0068
Axial Tilt or Obliquity	177.4°
(inclination of equator to orbit)	
Orbital Tilt	3.39°
(inclination of orbit to ecliptic)	
Magnetic Field (measured)	$< 5 \times 10^{-5}$ Gauss

7.3.3   Earth

Earth is the third planet from the Sun and the largest and most dense of the terrestrial planets. It is the fifth-largest planet in the Solar System and has approximately 70 percent of its surface covered with water.

Earth is the only planet known to support life and has an atmosphere composed primarily of nitrogen (78 percent by volume) and oxygen (21 percent by volume). Earth has a solid inner core of iron and nickel.

Rotation – It takes Earth one day to complete one spin on its axis. Because of its solid nature, Earth is a non-differential rotating body. Although angular speed is constant, linear speed varies with latitude. Linear speed decreases as one travels north or south of the equator. Given the equatorial speed (V_E) and angle of latitude (phi) ϕ, the linear speed at any given latitude (V_L) can be determined from the following equation:

$$V_L = V_E \cos |\phi|$$

where:

V_L = Rotational speed at a specific latitude

V_E = Rotational speed at the equator (a constant for any given body)

$\cos |\phi|$ = cosine of angle of latitude (in degrees)

Differential Rotation – As a result of having a gaseous nature, all of the jovian planets and the Sun exhibit differential rotation. The angular rotational speed of such bodies varies with latitude. This causes the body to have different rotational periods for varying latitudes. In essence, the entire body does not rotate in unison.

Non-Differential Rotation – For spherically shaped non-differential rotating bodies, **angular speed** of rotation is the same at all latitudes. This results in having a single period of rotation for the entire body. As an example, Earth's rotational period is approximately 24 hours everywhere. All of the terrestrial planets and Pluto are solid bodies that exhibit non-differential rotation.

The linear speed of rotation of a non-differential rotating body is dependent upon latitude. Linear rotational speed is fastest at the equator and slowest (zero) directly at the poles. For example, Earth's linear rotational speed is 1,036 mph at the equator and zero mph at the centers of the North and South Poles. All other latitudes have linear rotational speeds somewhere in between these two values.

Revolution – It takes Earth one year to complete one trip around the Sun. Earth reaches perihelion (closest approach to Sun) around January 3 and aphelion (farthest distance) around July 4.

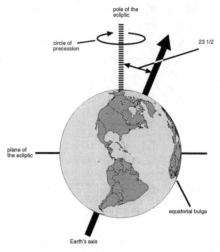

Figure 7.3   Precession

General Precession – This is the slow periodic change in the direction of the Earth's axis of rotation as a result of external torques by the Moon, Sun, and planets. One cycle of precession traces out a circle in the sky every 25,800 years. This motion results in continually changing pole stars and stellar coordinates.

Astronomical Nutation – Earth's axis of rotation nods like a spinning top, tracing out a slightly wavy circle around the ecliptic poles in a period of 18.6 years. This is primarily caused by external gravitational influences of the Moon as its orbit changes over this period. Nutation is superimposed on general precession.

Polar Motion – The Earth's axis of rotation (geographic poles) oscillates slightly over a period of 12 months as a result of seasonal changes in the distribution of mass on Earth's surface and in its atmosphere. In addition, movements of material within the Earth's interior cause a 14-month oscillation to occur called the Chandler Wobble. The combination of these two oscillations causes the geographic poles to trace a spiral path back to their starting positions in about 6.5 years. This motion causes a slight variation in latitude and longitude coordinates.

Solar Motions – Earth and all other members of the Solar System are being dragged through space by the Sun at a speed of 12 miles per second toward the bright star Vega in the constellation of Lyra. Earth, as part of the Solar System, is also orbiting the galactic center at a speed of 155 miles per second. One orbit takes about 220 million Earth years.

EARTH DATA CHART

Mass	5.98×10^{24} kg
Equatorial Diameter	12,756 km or
	7,927 mi
Volume	1 Earth Volume
Average Density	5.52 g/cm³
Equatorial Surface Gravity	1 Earth Gravity
Equatorial Escape Speed	11.2 km/s or
	6.97 mi/s
Average Surface Temperature	–60° F to 120° F
Natural Satellites	1
Ring System	none
Average Albedo	0.39
Maximum Distance from Sun	1.521×10^{8} km or
	9.451×10^{7} mi
Minimum Distance from Sun	1.471×10^{8} km or
	9.142×10^{7} mi
Average Distance from Sun	1.495×10^{8} km or
	9.290×10^{7} mi

Sidereal Rotational Period	23 hrs 56 min 04 sec
Synodic Rotational Period	24 hrs 00 min 00 sec
Oblateness	0.0034
Sidereal Orbital Period	1.0 yr (365.26 days)
Average Orbital Speed	29.8 km/s or
	18.5 mi/s
Orbital Eccentricity	0.0167
Axial Tilt or Obliquity	23.5°
(inclination of equator to orbit)	
Orbital Tilt	0°
(inclination of orbit to ecliptic)	
Magnetic Field (measured)	0.31 Gauss

7.3.4   Mars

Mars is the fourth and outermost terrestrial planet from the Sun. It is visible to the unaided eye as a reddish starlike object due to the iron oxide dust on its surface and in its atmosphere. The thin Martian atmosphere is 95 percent carbon dioxide by volume. The polar ice caps consist of frozen carbon dioxide and ice. Mars has a solid rocky core.

MARS DATA CHART

Mass	10.8% of Earth or
	6.424×10^{23} kg
Equatorial Diameter	6,794 km or
	4,222 mi
Angular Diameter	17.9 arcseconds
(at closest approach)	

Volume	15% of Earth
Average Density	3.94 g/cm^3
Equatorial Surface Gravity	37.9% of Earth
Equatorial Escape Speed	5.0 km/s or
	3.1 mi/s
Average Surface Temperature	−220° F to 68° F
Natural Satellites	2
Ring System	none
Maximum Apparent Magnitude	−2
Average Albedo	0.16
Maximum Distance from Sun	2.492 × 10^8 km or
	1.549 × 10^8 mi
Minimum Distance from Sun	2.066 × 10^8 km or
	1.284 × 10^8 mi
Average Distance from Sun	2.279 × 10^8 km or
	1.416 × 10^8 mi
Sidereal Rotational Period	24 hrs 37 min 23 sec
Oblateness	0.0052
Sidereal Orbital Period	1.88 yrs
Average Orbital Speed	24.1 km/s or
	14.98 mi/s
Orbital Eccentricity	0.0934
Axial Tilt or Obliquity	25.2°
(inclination of equator to orbit)	
Orbital Tilt	1.85°
(inclination of orbit to ecliptic)	
Magnetic Field (measured)	< 5 × 10^{-4} Gauss

7.4　Jovian Planets

Jupiter, Saturn, Uranus, and Neptune represent four of the five outermost planets of the Solar System. They are gaseous giants known as the Jovian (Jupiter-like) planets. All of them have ring systems.

7.4.1　Jupiter

Visible to the unaided eye, Jupiter is the largest and fifth planet from the Sun. The four largest natural satellites of Jupiter are called the Galilean satellites in honor of their discoverer, Galileo. Jupiter's Great Red Spot is an oval-shaped, huge storm system first observed in the year 1664. Atmospheric gases are primarily hydrogen (82 percent by mass) and helium (18 percent by mass). Jupiter has a rocky core.

JUPITER DATA CHART

Mass	1.899×10^{27} kg or
	318 times Earth
Equatorial Diameter	142,800 km or
	88,736 mi
Angular Diameter	46.9 arcseconds
(at closest approach)	
Volume	1,323 Earths
Average Density	1.34 g/cm^3
Gravity (at cloud base)	2.54 times Earth
Equatorial Escape Speed	60 km/s or
	37 mi/s
Temperature (at cloud tops)	$-180°$ F
Natural Satellites	16
Ring System	1 main ring

Maximum Apparent Magnitude	−2.8
Average Albedo	0.5
Maximum Distance from Sun	8.160×10^8 km or
	5.071×10^8 mi
Minimum Distance from Sun	7.406×10^8 km or
	4.602×10^8 mi
Average Distance from Sun	7.783×10^8 km or
	4.836×10^8 mi
Sidereal Rotational Period	9 hrs 50 min 30 sec
(differential rotation)	(<12° latitude)
Oblateness	0.0625
Sidereal Orbital Period	11.87 Earth yrs
Average Orbital Speed	13.1 km/s or
	8.1 mi/s
Orbital Eccentricity	0.0484
Axial Tilt or Obliquity	3.1°
(inclination of equator to orbit)	
Orbital Tilt	1.3°
(inclination of orbit to ecliptic)	
Magnetic Field (measured)	4.28 Gauss

7.4.2 Saturn

Saturn is the sixth planet from the Sun and is also visible to the unaided eye. Unlike the other Jovian planets, its ring system can be seen through a small telescope. Its atmosphere is principally hydrogen (93 percent by volume) and helium (3 percent by volume). Saturn has a core of rock and ice.

SATURN DATA CHART

Mass	5.69×10^{26} kg or
	95 times Earth
Equatorial Diameter	120,660 km or
	74,978 mi
Angular Diameter	19.5 arcseconds
(at closest approach)	
Volume	744 Earths
Average Density	0.69 g/cm^3
Gravity (at cloud base)	1.16 times Earth
Equatorial Escape Speed	35.6 km/s or
	22.1 mi/s
Temperature (at cloud tops)	−292° F
Natural Satellites	19 minimum
Ring System	7 major rings
Maximum Apparent Magnitude	+0.3
Average Albedo	0.61
Maximum Distance from Sun	1.507×10^9 km or
	9.365×10^8 mi
Minimum Distance from Sun	1.347×10^9 km or
	6.370×10^8 mi
Average Distance from Sun	1.427×10^9 km or
	8.867×10^8
Sidereal Rotational Period	10 hrs 13 min 59 sec
(differential rotation)	(near equator)

Oblateness	0.1087
Sidereal Orbital Period	29.46 Earth yrs
Average Orbital Speed	9.64 km/s or
	6 mi/s
Orbital Eccentricity	0.0560
Axial Tilt or Obliquity	26.7°
(inclination of equator to orbit)	
Orbital Tilt	2.49°
(inclination of orbit to ecliptic)	
Magnetic Field (measured)	0.021 Gauss

7.4.3   Uranus

Uranus is the seventh planet from the Sun and exhibits a bluish-green color due to its methane gas. Atmospheric gases are principally hydrogen (85 percent by volume) and helium (15 percent by volume). Uranus has a solid rocky core.

URANUS DATA CHART

Mass	8.69×10^{25} kg or
	14.5 times Earth
Equatorial Diameter	51,118 km or
	31,765 mi
Angular Diameter	3.6 arcseconds
(at closest approach)	
Volume	67 Earths
Average Density	1.19 g/cm^3
Equatorial Surface Gravity	91% of Earth

Equatorial Escape Speed	21 km/s or
	13 mi/s
Temperature (at cloud tops)	–366° F
Natural Satellites	15
Ring System	11
Maximum Apparent Magnitude	+5.6
Average Albedo	0.35
Maximum Distance from Sun	3.00×10^9 km or
	1.87×10^9 mi
Minimum Distance from Sun	2.74×10^9 km or
	1.70×10^9 mi
Average Distance from Sun	2.869×10^9 km or
	1.783×10^9 mi
Sidereal Rotational Period	17 hrs 54 min (retrograde)
(differential rotation)	at Equator
Oblateness	0.024
Sidereal Orbital Period	84 yrs
Average Orbital Speed	6.8 km/s or
	4.2 mi/s
Orbital Eccentricity	0.0461
Axial Tilt or Obliquity	98°
(inclination of equator to orbit)	
Orbital Tilt	0.774°
(inclination of orbit to ecliptic)	
Magnetic Field (measured)	0.23 Gauss

7.4.4 Neptune

Neptune is the eighth planet from the Sun and exhibits a bluish color as a result of its methane content. Its atmosphere is principally hydrogen (85 percent by volume) and helium (13 percent by volume). The oval-shaped feature, called the Great Dark Spot, is a huge storm system. Neptune has a rocky silicate core.

NEPTUNE DATA CHART

Mass	1.030×10^{26} kg or
	17 Earth masses
Equatorial Diameter	49,500 km or
	30,759 mi
Angular Diameter	2.12 arcseconds
(at closest approach)	
Volume	57 Earths
Average Density	1.66 g/cm^3
Equatorial Surface Gravity	1.2 times Earth
Equatorial Escape Speed	24 km/s or
	14.9 mi/s
Temperature (at cloud tops)	$-357°$ F
Natural Satellites	8
Ring System	4
Maximum Apparent Magnitude	+7.9
Average Albedo	0.62
Maximum Distance from Sun	4.54×10^9 km or
	2.82×10^9 mi
Minimum Distance from Sun	4.452×10^9 km or
	2.766×10^9 mi

Average Distance from Sun	4.497×10^9 km or
	2.795×10^9 mi
Sidereal Rotational Period	17 hrs 48 min
(differential rotation)	(at midlatitudes)
Oblateness	0.027
Sidereal Orbital Period	164.79 Earth yrs
Average Orbital Speed	5.4 km/s or
	3.4 mi/s
Orbital Eccentricity	0.0100
Axial Tilt or Obliquity	28.8°
(inclination of equator to orbit)	
Orbital Tilt	1.774°
(inclination of orbit to ecliptic)	
Magnetic Field (measured)	0.1 Gauss

7.4.5   Pluto

Pluto cannot be classified as a jovian or terrestrial planet as its composition is different. Pluto is usually the farthest planet from the Sun, but its highly eccentric orbit causes it to pass inside of Neptune's orbit. This results in Neptune being farther from the Sun for 20 years of Pluto's 248 year orbit around the Sun. Pluto's rotation is retrograde. Its atmosphere is probably methane and its core is probably rock and ice.

PLUTO DATA CHART

Mass	1.2×10^{22} kg or
	0.2% of Earth
Equatorial Diameter	2,294 km or
	1,425 mi

Angular Diameter (at closest approach)	0.08 arcseconds
Volume	unknown
Average Density	1.64 g/cm^3
Equatorial Surface Gravity	6% of Earth's
Equatorial Escape Speed	1.2 km/s or 0.75 mi/s
Surface Temperature	−382° F
Natural Satellites	1
Ring System	none
Maximum Apparent Magnitude	13.6
Average Albedo	0.4
Maximum Distance from Sun	7.366 × 10^9 km or 4.557 × 10^9 mi
Minimum Distance from Sun	4.434 × 10^9 km or 2.755 × 10^9 mi
Average Distance from Sun	5.900 × 10^9 km or 3.666 × 10^9 mi
Sidereal Rotational Period	6 days 9 hrs 21 min (retrograde)
Oblateness	unknown
Sidereal Orbital Period	247.7 Earth years
Average Orbital Speed	4.73 km/s or 2.94 mi/s
Orbital Eccentricity	0.2484

Axial Tilt or Obliquity	118° (estimated)
(inclination of equator to orbit)	
Orbital Tilt	17.2°
(inclination of orbit to ecliptic)	
Magnetic Field	unknown

7.5   Titius-Bode Rule

The Titius-Bode rule is a mathematical expression that approximates the distances of the planets from the Sun. It predicts a planet at a distance of 2.8 A.U.'s (the asteroid belt) and is not accurate for the two outermost planets, Neptune and Pluto. The formula is as follows:

$$D = 0.4 + (0.3 \times N)$$

where:

D = Distance (in A.U.'s)

N = The values 0, 1, 2, 4, 8, 16, 32, 64, 128 and 256 which correspond to the planets and asteroid belt in order of increasing distance from the Sun.

CHAPTER 8

Planetary Satellites

8.1 Artificial Satellites

An artificial satellite is a man-made object placed in orbit around Earth or some other celestial body. These satellites are designed to perform various functions (e.g., communication links, weather information, and gather research data).

8.2 Natural Satellites

A natural satellite is a body of natural origin in orbit around a planet. Except for Mercury and Venus, all planets have natural satellites. Natural satellites are commonly referred to as "moons." All natural satellites have proper names. Earth's natural satellite is named Moon.

8.3 The Moon

Second to the Sun in apparent brightness, the Moon is Earth's nearest neighbor and only natural satellite. Its surface is easily seen as highlands (heavy cratered regions) and lowlands (smooth, dark plains called marias). The Moon has no atmosphere or water and has a small inner core that is probably molten rock or iron. On the average, the Moon rises about 50 minutes later each successive day.

8.4   Moon Phases

The Moon goes through a series of phases every 29½ days (synodic period). Depending on where the Moon is in its orbit around Earth, we see differing amounts of its sunlit side.

After 19 tropical years (365.24219 days) the cycle of Moon phases recur on the same days of the year. This is known as the Metonic cycle.

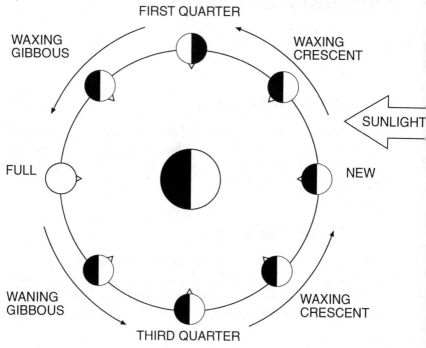

Figure 8.1   Phases of the Moon

8.5   Moon Motions

Besides spinning on its axis (rotating) and orbiting (revolving) around Earth, the Moon exhibits a motion called libration.

8.5.1 Rotation

The Moon is in a synchronous, or captured, rotation. That is, its rotational period matches its orbital period. With respect to the Sun, this period (a synodic, or lunar month) is 29½ days; with respect to the background stars, it is 27⅓ days. This results in the Moon always having its same side facing Earth.

8.5.2 Revolution and Libration

As the Moon orbits Earth, it exhibits librations of latitude and longitude. As a result of these motions, we see more of the Moon's polar regions and east and west areas. Rather than seeing just 50 percent of the Moon's surface, we can see 59 percent.

8.5.3 Tides

Tides are the rising and falling of Earth's waters and, to a lesser extent, its land masses as a result of gravitational effects from the Moon and Sun. Although far less massive than the Sun, the Moon's closer distance causes it to have about twice the tidal force upon Earth.

Spring tides are high tides and occur because the Moon and Sun are positioned in line with Earth so that their gravitational fields reinforce each other. This alignment only happens during new Moon or full Moon. During this time, ocean water levels are at their highest.

Figure 8.2   Spring Tides

Neap tides are low tides and occur because the Moon and Sun are positioned at right angles to each other with respect to Earth. This alignment, which can only occur during first quarter Moon and last quarter Moon, causes the gravitational fields of the Earth and Sun to work at right angles to each other. During this time, ocean water levels are at their lowest.

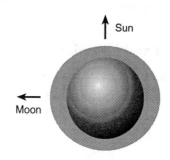

Figure 8.3   Neap Tides

8.5.4   Moon Data Chart

Mass	7.35×10^{22} kg or
	1.2% of Earth
Equatorial Diameter	3,476 km or
	2,160 mi
Angular Diameter	29.5 to 32.9 arcminutes
Volume	2% of Earth
Average Density	3.36 g/cm^3
Equatorial Surface Gravity	16.7% of Earth
Equatorial Escape Speed	2.38 km/s or
	1.48 mi/s
Average Surface Temperature	$-274°$ F to 266° F
Ring System	none

Maximum Apparent Magnitude	−12.5 (full Moon)
Absolute Albedo	0.07
Maximum Distance from Earth	405,500 km or
	251,978 mi
Minimum Distance from Earth	363,300 km or
	225,755 mi
Average Distance from Earth	384,400 km or
	238,866 mi
Sidereal Rotational Period	27⅓ days
Synodic Rotational Period	29½ days
Oblateness	0
Sidereal Orbital Period	27.3 days
Synodic Orbital Period	29.5 days
Average Orbital Speed	3,680 km/hr or
	2,287 mi/hr
Orbital Eccentricity	0.055
Axial Tilt or Obliquity	6.68°
(inclination of equator to orbit)	
Orbital Tilt	5.15°
(inclination of orbit to ecliptic)	
Magnetic Field	none

CHAPTER 9

Asteroids, Meteoroids, and Comets

9.1 General Information

Asteroids, meteoroids, and comets are debris left over from the birth of the Solar System.

9.1.1 Asteroids

Asteroids are minor planets, also called planetoids. These solid rocky bodies are remnants of the solar nebula. They vary widely in size and shape. Ceres is the largest known asteroid and has a diameter of approximately 1,000 km (620 miles).

9.1.2 Asteroid Belt

Thousands of asteroids orbit the Sun between the orbits of Mars and Jupiter. This sparsely occupied region is known as the asteroid belt. The orbital periods of these asteroids range from three to six years. The asteroid belt marks the transition zone between the inner and outer planets.

9.1.3　Meteoroids

A meteoroid is any small solid celestial object moving through space having the potential of becoming a meteor or meteorite.

9.1.4　Meteors

A meteor is a meteoroid that has entered Earth's atmosphere. Due to friction, the meteor heats up and begins to glow while streaking across the sky. Meteors are commonly called shooting stars and falling stars.

An unusually bright meteor is called a fireball and a fireball that explodes along its path is called a bolide.

Meteor showers occur when Earth passes through debris generally left behind by orbiting comets. Debris left behind by comet Swift-Tuttle is the source of meteoric material for the annual Perseids meteor shower.

9.1.5　Meteorites

A meteorite is a meteor that survives passage through Earth's atmosphere and reaches its surface. There are three main classes of meteorite: irons (siderites), stony-irons (siderolites or lithosiderites) and stones (aerolites). Stones are the most common types of meteorites seen to fall and are divided into two classes; chondrites (most common type of stony meteorites) and achondrites.

9.2　Comets

A comet is an icy celestial body orbiting in the Solar System. As it approaches the Sun, it partially vaporizes and forms a diffuse envelope of gas around the nucleus; this is called the coma. The nucleus is a mixture of dust and frozen gases that consist primarily of water, carbon dioxide, ammonia, and methane. The coma and nucleus form the head of the comet where long trails of gas and dust stream away to form the dust and gas tails. As Earth orbits the Sun, it will periodically encounter debris left behind by comets. The passage of Earth through this debris is what causes meteor showers.

9.2.1 Periodic Comets

These are comets in a closed elliptical orbit within the Solar System. They have either short periods (6 years to 200 years) or long periods (thousands of years). Halley's comet is a short-period comet with a period of about 76 years.

9.2.2 Non-Periodic Comets

These comets are merely visitors passing through the Solar System. Since they move along open orbits that are hyperbolic, they can never return.

CHAPTER 10

Eclipses

10.1 Definition

An eclipse occurs when light from a celestial body is partially or completely blocked by the presence of another. Nearly identical cycles of solar and lunar eclipses repeat after approximately 18 years 11⅓ days. This is known as a Saros period. Subsequent eclipse sequences will occur approximately eight hours later and 120° farther west.

10.2 Solar Eclipse

A solar eclipse occurs when the Moon lies between the Sun and the Earth, thereby blocking the Sun's light. Solar eclipses can only occur at or near new Moon.

10.2.1 Total Solar Eclipse

A total solar eclipse occurs when the Moon's disk completely covers the Sun's. Although the Sun's diameter is about 400 times greater than the Moon's, its distance from Earth is about 400 times greater. As a result, both bodies have the same angular diameter of about one-half degree (30 arcseconds).

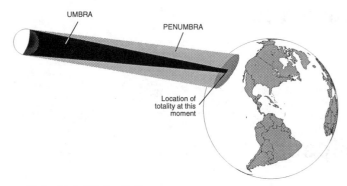

Figure 10.1 Total Solar Eclipse

10.2.2 Partial Solar Eclipse

A partial solar eclipse occurs when the Moon's disk partially covers the Sun's.

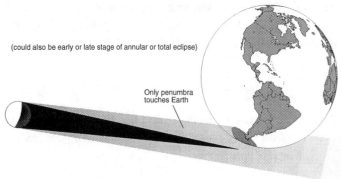

Figure 10.2 Partial Solar Eclipse

10.2.3 Annular Eclipse

An annular eclipse occurs when the Moon is at or near apogee (farthest from Earth). The Moon's angular diameter now appears smaller than the Sun's and thus fails to completely cover it. This results in a ring of sunlight surrounding the Moon.

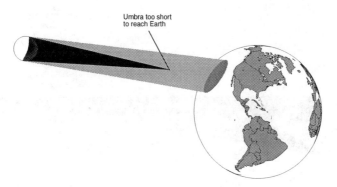

Umbra too short
to reach Earth

Figure 10.3   Annular Solar Eclipse

10.3   Lunar Eclipse

A lunar eclipse occurs when the Moon enters the Earth's shadow. The Earth's shadow then wholly or partially blocks the Moon's light. Lunar eclipses can only occur at or near full Moon.

10.3.1   Total Lunar Eclipse

When the Moon lies wholly within Earth's umbra (the dark central part of its shadow), the eclipse is total and the Moon's disk is dark (see figure 19).

10.3.2   Partial Lunar Eclipse

When the Moon lies partially in the umbra and penumbra (the lighter outer part of Earth's shadow), the eclipse is partial and only part of the Moon's disk is dark.

10.3.3   Penumbral Eclipse

When the Moon lies wholly within Earth's penumbra, the eclipse is penumbral. This type of eclipse can easily go unnoticed as the Moon's disk is only slightly darkened.

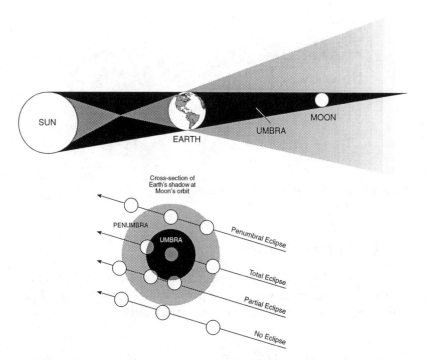

Figure 10.4   Penumbral Eclipse

10.4   Relationship Between Angular Diameter and Linear Diameter (for Small Angles)

$$A = \frac{L}{D}(206,265 \text{ arc seconds})$$

where:

A = Angular diameter of body (in arcseconds)

L = Linear diameter of body (in linear units)

D = Separation distance between Earth and body (in same units as "L")

CHAPTER 11

Stars

11.1   What is a Star?

A star is a luminous ball of hot ionized gas (plasma), generating energy through thermonuclear fusion. Stars more massive than the Sun generate their thermonuclear energy by converting three helium nuclei (alpha particles) into one Carbon-12 nucleus. This is known as the triple alpha process and is responsible for building elements heavier than helium. The process of building elements is called nucleosynthesis. In stellar interiors, this so-called "nuclear burning" proceeds with the successive burning of hydrogen, helium, carbon, neon, oxygen and silicon; iron is then created, halting this series. Elements more massive than iron are produced inside of red giants and supernovae by other nucleosynthetic processes. Stars are born from huge interstellar clouds of gas and dust called nebulae. We obtain information on stars by analyzing their light through spectroscopy. Spectral analysis reveals information that includes temperature, composition, and relative motion. Stars vary widely in mass, diameter, temperature, and many other properties. For example, a neutron star is about the size of a small city and a super giant star placed at the Sun's position would reach to the outer planets.

All of the stars seen in the sky are part of our Milky Way Galaxy. As seen from the North and South Poles, stars circle the horizon, never rising or setting. Moving towards the equator, they begin to rise and set at increasing angles until they rise and set at right angles to the equator's horizon. Stars rise about four minutes earlier each successive night and appear to twinkle because of atmospheric turbulence. They are always in the sky, but their presence is not always made known. During daylight hours, the brightness of the Sun makes it impossible to see the stars; at night, they may be hidden by clouds and/or light pollution.

11.2 Binary Star Systems

A true binary star is a pair of stars in orbit around each other and bound together by their mutual gravitational attraction. They are also called double stars. The classification of binaries includes those stars that appear to have nearby companions.

11.2.1 Optical Double

An optical double is a pair of stars that lie in the same line of sight as seen from Earth. They are not physically associated with each other and, therefore, do not constitute a true binary. The star Mizar, second in the handle of the Big Dipper, along with its companion Alcor, is an example.

11.2.2 Visual Binary

A visual binary is a star in which the two components can be resolved as separate images with an appropriately sized telescope.

11.2.3 Composite Spectrum Binary

A composite spectrum binary is a pair of stars that are resolved by observing lines of two different spectral types.

11.2.4 Eclipsing Binary

As seen from Earth, an eclipsing binary is a star system in which the orbital motion of the two stars causes them to pass in front of each other and thus reduce their light. This fluctuation in total brightness varies on a regular cycle. The star Algol, in the constellation of Perseus, is a classical example.

11.2.5 Astrometric Binary

An astrometric binary is a star in which the presence of an unseen companion is revealed by cyclic irregularities in the position of the brighter star.

11.2.6 Spectroscopic Binary

A spectroscopic binary is a system in which the two stars are so close together that the nature of their presence must be revealed by spectroscopic analysis. **Single-lined spectroscopic binaries** only reveal the spectrum of a brighter companion while **double-lined spectroscopic binaries** reveal the spectra of both stars.

11.3 Multiple Star Systems

A multiple-star system is a group of three or more stars orbiting each other and bound together by their mutual gravitational attraction. It is estimated that at least one-half of all stars are either binary or multiple. The North Star, Polaris, and the bright star, Alpha Centauri, are both part of two different triple star systems.

11.4 Variable Stars

A variable star is any star whose light output varies regularly or irregularly. This variation of light output is caused by various factors. Internal instability could be one factor, and the periodic passing of one star in front of the other could be another. There are three major groups of variable stars: eruptive and cataclysmic variables, pulsating variables, and eclipsing binaries. Many types of stars are classified as variable. They include Novae, Supernovae, T Tauri stars, Flare stars, Cepheid variables, RR Lyrae stars, Mira stars, and BY Draconis stars.

11.4.1 Cepheid Variables

A Cepheid variable is a luminous yellow giant star that changes regularly in brightness. This star physically oscillates due to an unstable structure. It was discovered that the brightest Cepheids have the longest periods. This relationship is known as the period-luminosity relationship and allows Cepheids to be used as distance indicators. Cepheids are pulsating variables. The North Star, Polaris, is a Cepheid variable.

11.4.2 Novae

A nova is a binary star system that suddenly increases its brightness by as much as ten magnitudes (almost 10,000 times) and then slowly decreases over a period of months. This is caused by a white dwarf accreting (amassing) material from a normal companion star. Novae are eruptive variables.

11.4.3 Supernovae

A supernova is a catastrophic stellar explosion releasing such enormous amounts of energy that the exploding star outshines an entire galaxy of billions of stars. The Crab Nebula, located in the constellation of Taurus, is a supernova remnant first observed in 1054 A.D. Its core is a pulsar. There are two main types of supernovae: Type I, the brightest, and Type II, the better understood. Type I supernovae are subdivided into Types Ia and Ib. Supernovae are cataclysmic variables.

11.5 Light

Light is that part of the electromagnetic spectrum that can be seen by the human eye. As with all electromagnetic radiation, it exhibits a dual wave-particle nature and propagates through space at a speed of 3×10^8 meters per second. Visible white light is composed of the seven colors of the rainbow. They are red, orange, yellow, green, blue, indigo and violet. The entire spectrum ranges from long wavelength-low energy and low frequency radio waves—to short wavelength-high energy and high frequency cosmic photons. The following are designated regions of the electromagnetic spectrum from least to most energetic: radio wave, radar wave, microwave, infrared, visible light, ultraviolet, x-ray, gamma ray and cosmic photon. A theoretical object that absorbs all of the radiation incident on it is called a "black body." It also behaves as a perfect radiator emitting radiation as predicted by Planck's law. Stars radiate energy very similar to black bodies.

11.5.1 Spectra

A spectrum results when a beam of electromagnetic radiation is dispersed into its component wavelengths. This can be accomplished with either a prism or diffraction grating. The three principal types of spectra are continuous, emission line, and absorption line.

According to Kirchhoff's first rule, a continuous spectrum will result when a solid, liquid or dense gas is heated. The filament of an incandescent bulb produces this type of spectrum.

According to Kirchhoff's second rule, an emission line spectrum will result when a hot, low density (transparent) gas produces light. This is also called a bright line spectrum. A neon sign produces this type of spectrum.

According to Kirchhoff's third rule, an absorption line spectrum will result when light from a continuous spectrum passes through a gas at a lower temperature. The atoms in this cooler, low density gas absorb certain wavelengths of the light. Light from the Sun and other stars produces this type of spectrum.

11.6 Magnitudes

The Stellar Magnitude scale is a quantitative measure of the brightness of stars and all other celestial objects. It follows a logarithmic scale where each successive increase is about 2.5 times brighter. On this scale, the lowest numbers correspond to the brightest objects (e.g., magnitude 0 is brighter than magnitude +1, but fainter than magnitude

–1). The faintest magnitude detectable by the unaided human eye is +6. The Sun at apparent magnitude –27 is the brightest star in the entire sky, while Sirius (the dog star) at apparent magnitude –1.5 is the brightest in the night sky.

11.6.1 Absolute Magnitude

Absolute magnitude is a measure of an object's brightness as seen from an arbitrary reference distance of 10 parsecs (.32.6 light years). Absolute magnitude depends upon intrinsic luminosity.

11.6.2 Apparent Magnitude

Apparent magnitude is a measure of an object's brightness as seen from Earth. Apparent magnitude depends upon intrinsic luminosity and distance. At a distance of 10 parsecs from Earth, apparent magnitude equals absolute magnitude.

11.6.3 Mathematical Relationship Between Absolute and Apparent Magnitudes

$$M = m - 5 \log_{10}\left(\frac{d}{10\,pc}\right)$$

where:

M = Absolute magnitude (on stellar magnitude scale)

m = Apparent magnitude (on stellar magnitude scale)

d = Separation distance (in parsecs)

11.7 Stellar Classifications

Stars are classified in two ways: with respect to their spectrum and with respect to their luminosity.

11.7.1 Spectral Classification

The spectral classification system uses an alphabetical notation. It is easily recalled by reciting the mnemonic O̲h B̲e A̲ F̲ine G̲irl (Guy) K̲iss M̲e. The first letter of each word represents a main classification with temperatures decreasing from about 30,000 K (class O) to 3,500 K (class M). Each main class is further divided into 10 subgroups indicated by numbers (0 through 9). Lettered prefixes and suffixes are also used to indicate additional information.

11.7.2 Luminosity Classification

The luminosity classification system groups stars according to their luminosity (energy radiated/time). This classification system uses an uppercase Roman numeral notation, as follows.

Class Ia represents luminous supergiants

Class Ib represents less luminous supergiants

Class II represents luminous giants

Class III represents normal giants

Class IV represents subgiants

Class V represents dwarfs (main sequence stars)

Class VI represents subdwarfs

Class VII represents white dwarfs

Using both classification systems, the Sun is a type G2 V star.

11.8 Hertzsprung-Russel (H-R) Diagram

The Hertzsprung-Russel diagram is a plot of a star's luminosity (e.g., absolute magnitude) against some measure of its temperature (e.g., spectral type). A star's position on this diagram is determined by its stage of evolution, but this has no bearing on its location in actual space. On this diagram, approximately 90 percent of all stars (Sun included) fall within a narrow region called the main-sequence.

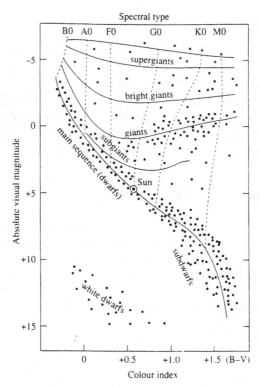

Figure 11.1   Hertzsprung-Russel Diagram

11.9   Determining Stellar Distances

Trigonometric parallax is the only direct method of determining distances to stars, but this method only works for those stars that are nearby. Since the vast majority of stars are too distant to have their parallaxes measured, indirect or statistical methods are used. The Sun is the closest star to Earth, and the next closest star is Proxima Centauri at a distance of about four light years. Many stars within our galaxy lie hundreds and even thousands of light years away. Stars are also separated from one another by light-year distances. The most distant object that can be seen with the unaided eye is the Andromeda galaxy at a distance of 2.2 million light years.

11.10 Constellations

Constellations are imaginary star patterns that divide the sky into specific regions. There are 88 such patterns in the entire sky surrounding Earth. The largest constellation is Hydra (1,303 sq. deg.) and the smallest is the Southern Cross (68 sq. deg.). Constellations do not represent the actual positions of stars in space. These star groupings appear connected because, as seen from Earth, the stars lie in that specific direction. Constellations are divided into subgroups called asterisms. The Big Dipper is an asterism in the constellation Ursa Major (Big Bear). Many of the constellations in use today originated in Mesopotamia and were further developed by the Greeks.

11.11 Stellar Nomenclature

One method of naming a star involves affixing a Greek letter before the name of the constellation (possessive case) in which the star lies. Greek letters can be assigned in three different ways. One way is based upon the natural ordering, or arrangement, of the stars within the pattern. Another way is based upon right ascension coordinates. And finally, assignment can be based upon the relative brightness of each star. The brightest star is assigned the first letter of the Greek alphabet (alpha), then the next brightest star is assigned the second letter (beta), and so on. Thus, Alpha Canis Majoris is the name of the brightest star in the constellation Canis Major.

Another method of naming stars involves the use of proper names. Only the brightest stars are called by such names. An example is the bright star "Sirius" (also known as Alpha Canis Majoris).

11.12 Stellar Motions

The motions of stars include proper motion, radial velocity, tangential velocity, spatial velocity, and peculiar velocity.

11.12.1 Proper Motion

Proper motion is the apparent angular shift of a star across the sky over a specified time period. This motion is actually a result of the star's true motion through space and the relative motion of the Solar System. Because of the proper motion of stars, constellations are slowly changing their shapes. Sirius has a proper motion of 1.3 arc seconds per year. The star with the largest proper motion is Barnard's star at 10.25 arc seconds per year.

11.12.2 Radial Velocity

This is the velocity in the direction along the line of sight. It describes the velocity of a star towards or away from the Sun. The radial velocity is positive if the star is moving away from the Sun and negative if it's approaching. This velocity can be determined from the Doppler shift.

11.12.3 Tangential Velocity

Also called transverse velocity, this is a star's velocity at right angles to an observer's line of sight.

11.12.4 Space Velocity

Space velocity is a star's total velocity relative to the Sun. It is a combination of the radial and tangential velocities.

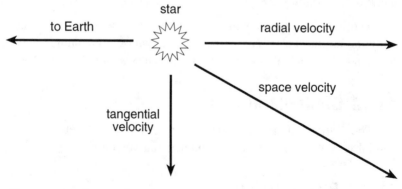

Figure 11.2   Space Velocity

11.12.5 Peculiar Velocity

Peculiar velocity is a star's velocity relative to a volume of space around the Sun where the space velocity of all the stars average out to zero. This reference area is called the local standard of rest but of course, is not really at rest.

11.13 Destiny of Stars

Stars are born from huge interstellar clouds of gas and dust, and their life expectancy depends on how massive they are. The more massive the star, the shorter its lifetime and the more violent its death.

11.13.1 White Dwarfs, Red Dwarfs, and Medium-Mass Stars

The white dwarf phase signals the end of life for low massive stars. Collapse of these stars is halted by the pressures of electron degeneracy. These low-mass stars are subdivided into two groups. One subgroup is "very low-mass" stars, such as red dwarfs; and the other is the "medium-mass" stars, such as the Sun. One teaspoon from these very dense stars would weigh about five tons.

White dwarfs have an upper mass limit of 1.4 solar masses. This is called the **Chandrasekhar limit**. Above this limit, a star collapses further to become either a neutron star or a black hole.

Red dwarfs are the most numerous class of stars in space. They are fainter and cooler than all of their fellow dwarf (main-sequence) stars. Most of their energy is emitted in the infrared, and they have very long nuclear lifetimes. A typical red dwarf might remain on the main-sequence, fusing hydrogen into helium for 100 billion years. Self-gravitation causes these very low-mass stars to contract, heat up-and move on to become white dwarfs. They gradually cool and dim out to become black dwarfs.

The Sun is a medium-mass star classified as a dwarf. It will spend 10 billion years of its life on the main-sequence, fusing hydrogen into helium. Hydrogen fusion will then cease, and the Sun's outer layers will begin to cool and expand. The Sun is now a red giant, fusing helium into carbon. After the Sun can no longer fuse helium, its outer layers will drift away to form an expanding gas shell called a planetary nebula. If the remaining core is less than 1.4 solar masses, it contracts to a very dense Earth-size star called a white dwarf. Ultimately, it cools and dims out to become a black dwarf.

11.13.2 Neutron Stars

Massive stars spend only millions of years on the main-sequence, fusing hydrogen into helium. Afterwards, they expand and become red super giants. For the following several million years, the building of elements (nucleosynthesis) occurs. This process halts and, in a fraction of a second, the core collapses and the outer layers are blown off in a massive explosion called a supernova. The subsequent luminosity can exceed that of the Sun several billion times. If there is a surviving core with a mass of no more than three solar masses, it will contract to neutron degeneracy and become a neutron star. These stars are extremely dense and small. One teaspoon of this stars material weighs about one billion tons. Neutron stars that are rapidly rotating emit consistent pulses of electromagnetic radiation and are called pulsars.

11.13.3 Black Holes

The ultimate fate for the most massive stars is a black hole. These are the collapsed cores of supernova remnants that had masses greater than three times that of the Sun. An implosion due to enormous gravitational forces causes the core to be literally crushed out of the visible universe. All that remains is an intense gravitational field so strong that nothing can escape from it, not even light. Black holes can be detected if they are part of a binary star system.

11.14 Masses of Binary Stars

$$M_1 + M_2 = \frac{a^3}{P^2}$$

where:

M_1 = Mass of star #1 (in solar masses)

M_2 = Mass of star #2 (in solar masses)

a = Average distance of separation (in AU's)

P = Orbital period (in years)

11.15 Escape Speed

$$\text{Escape Speed} = \sqrt{\frac{2GM}{R}}$$

where

G = Gravitational constant

M = Mass of body

R = Radius of body

11.16 Relationship Between Frequency and Wavelength (for Electromagnetic Radiation)

$$f = \frac{c}{\lambda}$$

where:

f = Frequency (in hertz or cycles per second)

λ = Wavelength (in meters)

c = Speed of light (3×10^8 m/s)

11.17 Relationship Between Energy and Wavelength (for Electromagnetic Radiation)

$$E = \frac{hc}{\lambda}$$

where:

E = Energy (in ergs)

λ = Wavelength (in centimeters)

h = Planck's constant (6.6262×10^{-27} erg/s)

c = Speed of light (3×10^{10} cm/s)

11.18 Temperature Scales and Conversion Formulas

	Temperature Scales		
Reference Points	**Celsius**	**Fahrenheit**	**Kelvin**
Boiling point of water	100°C	212°F	373K
Freezing point of water	0°C	32°F	273K
Absolute zero	−273.15°C	−459.69°F	0K

Conversion Formulas

$$°C = \frac{5}{9}(°F - 32)$$

$$°F = \left(\frac{9}{5}°C\right) + 32$$

$$K = °C + 273$$

11.19 Intensity Ratio of Two Bodies

$$\frac{I_1}{I_2} = (2.512)^{(M_1 - M_2)}$$

where:

I_1 = Intensity of the first body (energy rate per unit area)

I_2 = Intensity of the second body (energy rate per unit area)

M_1 = Magnitude of first body (on stellar magnitude scale)

M_2 = Magnitude of first body (on stellar magnitude scale)

11.20 Mass-Luminosity Relationship (for Main-Sequence Stars)

$$\frac{L}{L_S} = \left(\frac{M}{M_S}\right)^{\propto}$$

L = luminosity of star (in solar luminosities)

L_S = luminosity of star (in solar luminosities)

M = Mass of star (in solar masses)

M_S = Mass of sun (in solar masses)

$\propto$ = Exponent that varies with mass range

For low mass stars, $(M < 0.3M_S)$ $\propto$ = 1.8

For medium mass stars, $(0.3M_S < 3M_s)$ $\propto$ = 4.0

For high mass stars, $(M > 3M_S)$ $\propto$ = 2.8

11.21 Wien's Law (Used to Obtain the Temperature at the Surface of a Star)

$$T = \frac{2.9 \times 10^7}{\lambda_{max}}$$

where:

λ_{max} = Wavelength of maximum intensity (in Angstroms (Å) where
    1Å = 10^{-8} cm)

T = Temperature (in Kelvin)

11.22 Stefan-Boltzman Law (Used to Obtain the Energy or Luminosity of a Star)

$$E = \sigma T^4$$

where:

E = Total energy emitted per unit area

σ = Stefan-Boltzman constant $\left(5.672 \times 10^{-5} \dfrac{erg}{cm^2 K^4 s}\right)$

T = Temperature (in Kelvin)

11.23 Luminosity, Radius, and Temperature

$$L = 4\pi R^2 \sigma T^4$$

where:

L = Luminosity (in erg/s)

$4\pi R^2$ = Area of sphere (radius "R" in centimeters)

T = Temperature (in Kelvin)

σ = Stefan-Boltzman constant $\left(5.672 \times 10^{-5} \dfrac{erg}{cm^2 K^4 s}\right)$

11.23.1 Alternate Relationship for Luminosity, Radius, and Temperature (with Respect to the Sun)

$$\frac{L}{L_S} = \left(\frac{R}{R_S}\right)^2 \left(\frac{T}{T_S}\right)^4$$

where:

$\dfrac{L}{L_S}$ = Luminosity with respect to the Sun

$\dfrac{R}{R_S}$ = Radius with respect to the Sun

$\dfrac{T}{T_S}$ = Temperature with respect to the Sun

11.24 Doppler Shift and Radial Velocity

$$\frac{V_r}{c} = \frac{\Delta\lambda}{\lambda_o}$$

where:

V_r = Radial velocity (in km/s)

$\Delta\lambda$ = Change in wavelength (in nanometers = 10^{-9} meters)

λ_o = Unshifted wavelength (in nanometers)

c = Speed of light (3×10^5 km/s)

11.25 Life Expectancy of Stars on the Main-Sequence

$$T = \frac{1}{M_S^{\propto - 1}}$$

where:

T = Life expectancy of a main-sequence star (in solar lifetimes)

M_S = Mass of Sun (in solar masses)

$\propto$ = Exponent that varies with mass range

For low mass stars, $(M \leq 0.3M_S) \propto = 1.8$

For medium mass stars, $(0.3M_S < M \leq 3M_S) \propto = 4.0$

For high mass stars, $(M \geq 3M_S) \propto = 2.8$

11.26 Gravitational Red Shift (Especially Pertinent to Black Holes)

$$\frac{\lambda_2}{\lambda_1} = 1 + \frac{GM}{rc^2}$$

where:

$\dfrac{\lambda_2}{\lambda_1}$ = Gravitational red shift of wavelengths

G = Gravitational constant $\left(6.67 \times 10^{-11} \dfrac{N \times m^2}{kg^2} \right)$

M = Mass of body (in kilograms)

r = Radius of body (in meters)

c = Speed of light $(3 \times 10^8$ m/s$)$

11.27 Schwarzchild Radius

The Schwarzchild radius is the critical radius of a body that must be exceeded if light from its surface is to reach an outside observer.

$$R_S = \frac{2GM}{c^2}$$

where:

R_S = Schwarzchild radius (in meters)

G = Gravitational constant $\left(6.67 \times 10^{-11} \frac{N \times m^2}{kg^2} \right)$

M = Mass of the object (in kilograms)

c = Speed of light (3×10^8 m/s)

CHAPTER 12

The Observable Universe

12.1   Boundaries of the Observable Universe

On a large scale, the Universe looks the same in all directions and contains an even distribution of matter. These two assertions of isotrophy and homogeneity respectively describe what is known as the cosmological principle. The observable universe is that portion of all that exists within the confines of light travel time. Since light, and all other forms of electromagnetic radiation travel at an unsurpassable finite speed, we must patiently await their arrival before we can know what is, or once was.

12.2   The Big Bang

The Big Bang theory coupled with the Inflationary-Universe theory is a model attempting to explain how the Universe came to be as we now know it. Although it leaves many important questions unanswered, it is currently the most promising cosmological model. According to this model, the early Universe was very dense and very hot. From this condition, it began a violent expansion referred to as the Big Bang. Shortly thereafter, there was a brief period (between 10^{-45} seconds and 10^{-30} seconds) of incredible expansion during which the Universe inflated by a factor of billions. After this brief inflationary period, the Universe resumed its previous rate of expansion and cooling. This allowed for nucleosynthesis, the decoupling of radiation and matter (now seen as the cosmic background radiation), and the formation of protogalaxies and clusters. The advent of the inflationary period has resolved previously encountered cosmological problems. Additional support for this model includes the uniform expansion of the Universe, the cosmic background radiation, and the prediction of the abundance of light elements and isotopes. The Big Bang is believed to have had its beginning somewhere between 13 and 20 billion years ago. Hubble's constant (or parameter, as its value changes with time) is a numerical value that attempts to estimate the current rate at which the Universe is expanding with time. Current estimates place it somewhere between 50 and 100 kilometers per second per megaparsec (1 megaparsec = 10^6 pc).

The Big Bang model just described is also referred to as the "hot Big Bang." Other cosmological models include the "cold Big Bang" and the "Steady-State" model.

12.2.1 Estimating the Age of the Universe Using the Hubble Constant

$$T = \frac{1}{H_0}$$

where:

T = approximate age of universe or the Hubble Time

H_0 = Hubble's Constant or parameter (between 50 and 100 $\frac{km/s}{Mpc}$)

For upper age limit of Universe let $H_0 = 50\frac{km/s}{Mpc}$

then:

$$T = \frac{1}{\frac{50\,km/s}{Mpc}} = \frac{1}{50\,km \times sec^{-1}\,Mpc^{-1}} = \frac{1\sec \times Mpc}{50\,km}$$

and converting to years, we have

$$\frac{1\sec\times Mpc}{50\text{ km}}\times\frac{10^{6}pc}{Mpc}\times\frac{3.26\text{ Ly}}{1pc}\times\frac{6\times10^{12}\text{ mi}}{1\text{ Ly}}\times\frac{1\text{ km}}{0.6214\text{ mi}}\times$$

$$\frac{1\min_{x}}{60\sec}\times\frac{1\text{ hr}}{60\text{ min}}\times\frac{1\text{ day}}{24\text{ hr}}\times\frac{1\text{ year}}{365.24\text{ day}}\approx1.99\times10^{10}\text{ years}=$$

20 billion years (neglecting gravitational corrections).

12.3 Galaxies

A galaxy is a system of stars held together by their mutual gravitational attraction. The three main shapes or types are spiral, elliptical and irregular. The Sun is part of a spiral galaxy that consist of at least a few hundred billion stars. The name of this galaxy is the Milky Way, which is 100,000 light years across and 1,000 light years thick at the outer edges. Its central bulge has a radius of about 10,000 light years. The Sun is located approximately two-thirds of the way out from the center. Our galaxy is part of the "Local Group" of galaxies (a galactic cluster) that is suspected to contain at least 35 galaxies. The Local Group is part of a cluster of galactic clusters called a super cluster. Most galaxies are part of such a grouping and make up the Universe.

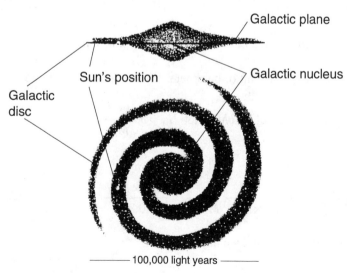

Figure 12.1 The Milky Way

12.4   Star Clusters

A star cluster is a group of stars, in close proximity, believed to have been formed from the same interstellar cloud. Star clusters are divided into two classes: open and globular.

12.4.1   Open Star Clusters

An open star cluster is a loosely scattered distribution of stars occupying a region a few light years across. Typically, these clusters contain anywhere from several hundred to several thousand stars. They are readily found in the disc of our galaxy. These are young clusters containing many hot and luminous stars. The Pleiades star cluster located in the constellation Taurus is an example of an open cluster.

12.4.2   Globular Clusters

A globular cluster is a somewhat spherically-shaped, densely-packed group of stars distributed within a spherical halo around the Galaxy. They can contain many thousands or even millions of stars. Some of the oldest stars (i.e., Population II) are found within these clusters. The globular cluster M13, in the constellation of Hercules, is the brightest in the Northern Hemisphere.

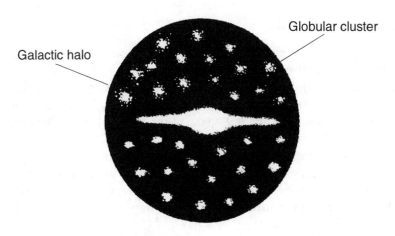

Figure 12.2   The Galactic Halo

12.5 Nebulae

A nebula is a cloud of interstellar gas and dust. Nebulae are classified as absorption, emission, and reflection.

12.5.1 Absorption Nebula

An absorption nebula, also called a dark nebula, absorbs the light from bright objects behind it. The Horsehead nebula, in the constellation Orion the Hunter, is an example.

12.5.2 Emission Nebula

An emission nebula, also called a bright nebula, glows as a result of bright stars that lie in it. The Great Orion Nebula, located in the sword of the constellation Orion the Hunter, is an example.

12.5.3 Reflection Nebula

A reflection nebula glows because its dust scatters light from nearby stars. The nebula surrounding the stars in the Pleiades open cluster is an example. This is also called a bright nebula.

12.6 Quasars

A quasar, also called a QSO, which is short for quasi-stellar object, is a small extragalactic object exhibiting a large red shift (as high as 4.9). A large red shift indicates a very high radial velocity of recession. For example, a red shift of 4.2 is equivalent to a radial velocity of about 93 percent the speed of light. Quasars have been observed at distances as far as 10 billion light years away. They are exceedingly bright for their small angular size of less than one arcsecond. There is high speculation that they may be the active nuclei of early galaxies (Active Galactic Nuclei). Their enormous source of energy may be generated by supermassive black holes.

12.7 Relativistic Red Shift and Radial Velocity

$$V_r = \frac{(Z+1)^2 - 1}{(Z+1)^2 + 1} \times c$$

where:

V_r = Radial velocity (in terms of c)

Z = Red shift = $\dfrac{\lambda - \lambda_0}{\lambda_0}$ = the change in wavelength divided by the unshifted wavelength.

c = Speed of light (in terms of c)

12.8 Hubble's Law

$$V_r = H_o d$$

where:

V_r = Radial velocity of galaxy (in km/s)

d = Distance to galaxy (in mega parsecs = Mpc)

H_o = Hubble's Constant = between 50 and 100 $\dfrac{\text{km/s}}{Mpc}$

CHAPTER 13

Extraterrestrial Intelligence

13.1 Life Outside of Earth

Extraterrestrial intelligence is any form of intelligent life that originates outside of the planet Earth. We have a vast universe in which to search, but our efforts seem somewhat hampered by the enormous distances separating the stars. The current research in this field focuses on radio signals from afar, biologically significant molecules in space, and the search for planetary systems around stars.

13.2 The Drake Equation

The Drake equation is one attempt at estimating our chances of finding and establishing communications with intelligent life elsewhere in the galaxy.

$$N = R_S f_p n_S f_L f_{int} f_{com} L$$

where:

R_S = Rate of stellar formation per galaxy

f_p = Fraction of stars per galaxy having planets

n_S = Number of planets per solar system suitable for life

f_L = Fraction of planets on which life develops

f_{int} = Fraction of these life forms that develop intelligence

f_{com} = Fraction of intelligent species that develop and use technology to communicate with other civilizations

N = Number of civilizations per galaxy having the ability to contact others within their galaxy

L = Average life expectancy of a communicative civilization

Based upon the estimated values from optimists, the Drake equation indicates the possible existence of a communicative civilization within several dozen light years from the Solar System. However, using the estimated values of pessimists yields a result indicating that we may be the only communicative civilization in our galaxy of a couple of hundred billion stars. Because of the high uncertainties of some of the estimated values of its parameters, a definitive answer to the Drake equation is at present inconclusive.

REA's **Problem Solvers**

The "PROBLEM SOLVERS" are comprehensive supplemental textbooks designed to save time in finding solutions to problems. Each "PROBLEM SOLVER" is the first of its kind ever produced in its field. It is the product of a massive effort to illustrate almost any imaginable problem in exceptional depth, detail, and clarity. Each problem is worked out in detail with a step-by-step solution, and the problems are arranged in order of complexity from elementary to advanced. Each book is fully indexed for locating problems rapidly.

ACCOUNTING
ADVANCED CALCULUS
ALGEBRA & TRIGONOMETRY
AUTOMATIC CONTROL
   SYSTEMS/ROBOTICS
BIOLOGY
BUSINESS, ACCOUNTING, & FINANCE
CALCULUS
CHEMISTRY
COMPLEX VARIABLES
DIFFERENTIAL EQUATIONS
ECONOMICS
ELECTRICAL MACHINES
ELECTRIC CIRCUITS
ELECTROMAGNETICS
ELECTRONIC COMMUNICATIONS
ELECTRONICS
FINITE & DISCRETE MATH
FLUID MECHANICS/DYNAMICS
GENETICS
GEOMETRY
HEAT TRANSFER

LINEAR ALGEBRA
MACHINE DESIGN
MATHEMATICS for ENGINEERS
MECHANICS
NUMERICAL ANALYSIS
OPERATIONS RESEARCH
OPTICS
ORGANIC CHEMISTRY
PHYSICAL CHEMISTRY
PHYSICS
PRE-CALCULUS
PROBABILITY
PSYCHOLOGY
STATISTICS
STRENGTH OF MATERIALS &
   MECHANICS OF SOLIDS
TECHNICAL DESIGN GRAPHICS
THERMODYNAMICS
TOPOLOGY
TRANSPORT PHENOMENA
VECTOR ANALYSIS

If you would like more information about any of these books,
complete the coupon below and return it to us or visit your local bookstore.

RESEARCH & EDUCATION ASSOCIATION
61 Ethel Road W. • Piscataway, New Jersey 08854
Phone: (732) 819-8880   **website: www.rea.com**

Please send me more information about your Problem Solver books

Name _____

Address _____

City _____ State _____ Zip _____

REA's Test Preps

The Best in Test Preparation

- REA "Test Preps" are **far more** comprehensive than any other test preparation series
- Each book contains up to **eight** full-length practice tests based on the most recent exam
- **Every** type of question likely to be given on the exams is included
- Answers are accompanied by **full** and **detailed** explanations

REA publishes over 60 Test Preparation volumes in several series. They include:

Advanced Placement Exams(APs)
Biology
Calculus AB & Calculus BC
Chemistry
Computer Science
Economics
English Language & Composition
English Literature & Composition
European History
Government & Politics
Physics B & C
Psychology
Spanish Language
Statistics
United States History

College-Level Examination Program (CLEP)
Analyzing and Interpreting
 Literature
College Algebra
Freshman College Composition
General Examinations
General Examinations Review
History of the United States I
History of the United States II
Human Growth and Development
Introductory Sociology
Principles of Marketing
Spanish

SAT II: Subject Tests
Biology E/M
Chemistry
English Language Proficiency Test
French
German

SAT II: Subject Tests (cont'd)
Literature
Mathematics Level IC, IIC
Physics
Spanish
United States History
Writing

Graduate Record Exams (GREs)
Biology
Chemistry
Computer Science
General
Literature in English
Mathematics
Physics
Psychology

ACT - ACT Assessment

ASVAB - Armed Services Vocational
 Aptitude Battery

CBEST - California Basic Educational
 Skills Test

CDL - Commercial Driver License Exam

CLAST - College Level Academic
 Skills Test

COOP & HSPT - Catholic High School
 Admission Tests

ELM - California State University Entry
 Level Mathematics Exam

FE (EIT) - Fundamentals of Engineering
 Exams - For both AM & PM Exams

FTCE - Florida Teacher Certification Exam

GED - High School Equivalency Diploma
 Exam (U.S. & Canadian editions)

GMAT CAT - Graduate Management
 Admission Test

LSAT - Law School Admission Test

MAT- Miller Analogies Test

MCAT - Medical College Admission Test

MTEL - Massachusetts Tests for
 Educator Licensure

MSAT- Multiple Subjects Assessment
 for Teachers

NJ HSPA - New Jersey High School
 Proficiency Assessment

NYSTCE: LAST & ATS-W - New York
 State Teacher Certification

PLT - Principles of Learning &
 Teaching Tests

PPST- Pre-Professional Skills Tests

PSAT - Preliminary Scholastic
 Assessment Test

SAT I - Reasoning Test

TExES - Texas Examinations of
 Educator Standards

THEA - Texas Higher Education
 Assessment

TOEFL - Test of English as a Foreign
 Language

TOEIC - Test of English for
 International Communication

USMLE Steps 1,2,3 - U.S. Medical
 Licensing Exams

U.S. Postal Exams 460 & 470

RESEARCH & EDUCATION ASSOCIATION
61 Ethel Road W. • Piscataway, New Jersey 08854
Phone: (732) 819-8880   **website: www.rea.com**

Please send me more information about your Test Prep books

Name _____

Address _____

City _____ State _____ Zip _____